Manhattan District History

Nonscientific Aspects of Los Alamos Project Y
1942 through 1946

Written by Edith C. Truslow and edited by Kasha V. Thayer

The Los Alamos Historical Society
Los Alamos, New Mexico
1991

ERRATA

Last paragraph, page 9, both acreage and dollar amount are incorrect. On May 6, 1943, the Declaration of Taking signed by Henry L. Stimson gave the Los Alamos Ranch School $275,000.00 for buildings, appurtenances, and 772.34 acres. The judgment filed on December 31, 1943, raised the compensation to $335,000.00 plus $7,884.98 in interest, making a total of $342,884.98.

A date in the appendix is incorrect. The year of Edward Wilder's arrival at Los Alamos was 1945, not 1944.

The few misspelled words throughout the original text were not corrected in this reprint.

Reprinted with permission of Los Alamos National Laboratory.
This document was originally issued in March 1973 as LA-5200,
based on an unpublished volume written in 1946.

Library of Congress Cataloging-in-Publication Data

Truslow, Edith C.
Manhattan District history : nonscientific aspects of Los Alamos Project Y, 1942 through 1946 / written by Edith C. Truslow and edited by Kasha V. Thayer.
p. cm.
ISBN 0-941232-11-5 (pbk.)
1. Manhattan Project (U.S.)—History. 2. United States. Army. Corps of Engineers. Manhattan District.—History. 3. Los Alamos Scientific Laboratory—History. I. Thayer, Kasha V. II. Title.
QC773.3.U5T78 1991
355.8'25119'0923—dc20 91-8207
CIP

Printed in the United States of America

The Trinity test, July 16, 1945.

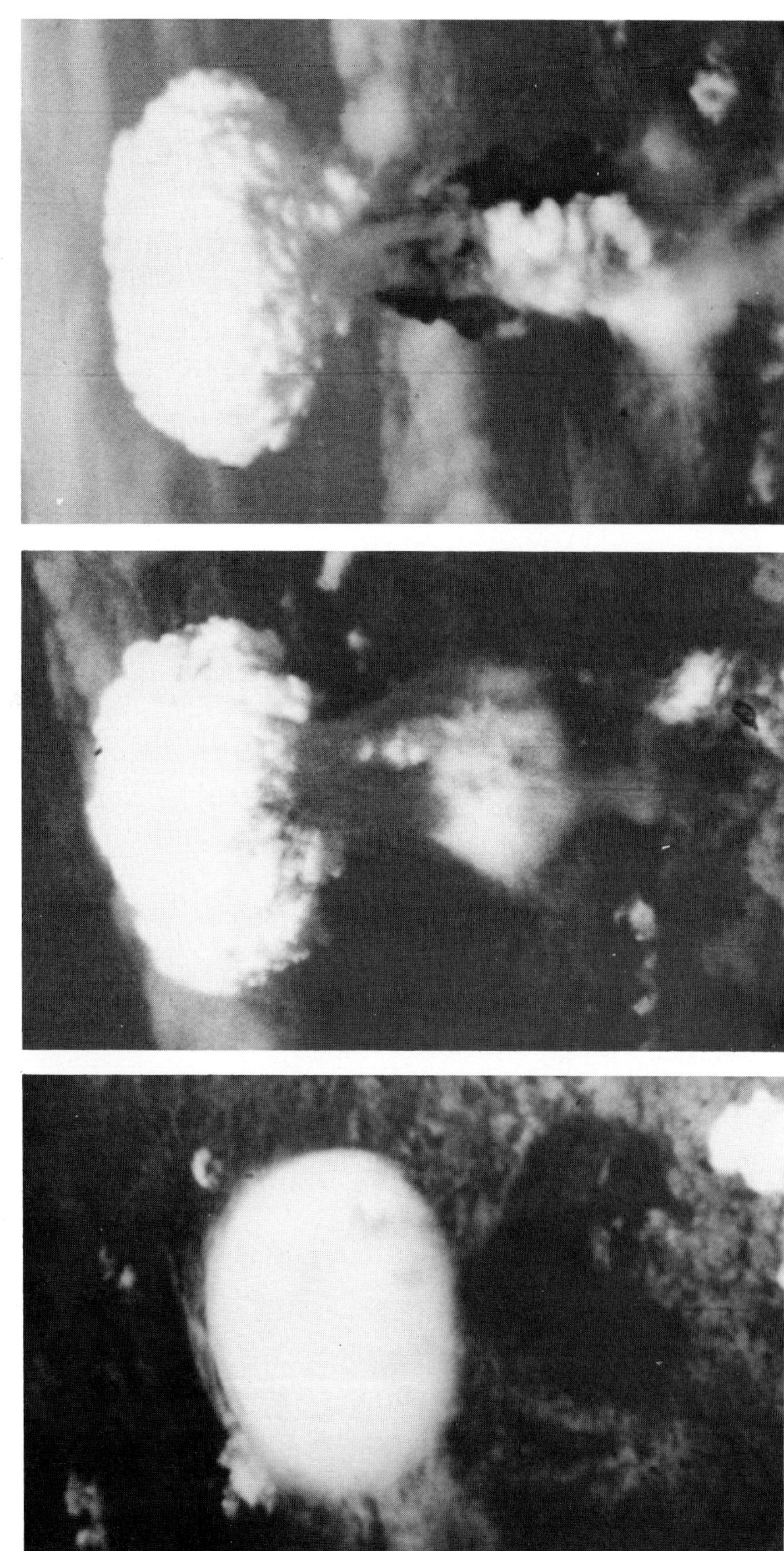

Nagasaki, August 9, 1945, from photographs taken by Harold M. Agnew.

Nagasaki.

FOREWORD

This volume of the Manhattan Engineer District History, by Edith C. Truslow who was a WAC 2nd Lietenant at the Project, gives a concise account of the nonscientific part of the Y Project at Los Alamos, New Mexico, from its inception through 1946, when the Atomic Energy Commission assumed control.

The ultimate achievements performed at Project Y are well known. Little can be said to add to this page of world history. The purpose of this record is to present the facts of construction, organization, and personnel and of various problems faced and met by those responsible for assisting in the final accomplishment of the task set for Los Alamos.

One of the chief difficulties in securing data for this volume arose from security restrictions during the early operations of the Project. Many of the transactions were carried on orally to ensure secrecy, and so no written record was preserved. In other instances, data were destroyed to further protect the secret of the bomb. This complete secrecy was one of the most amazing aspects of the entire program, but it produced one of the greatest obstacles to piecing the story together for historical purposes.

Fortunately, many of those involved in the founding of Los Alamos were still at the Project in 1946 and could supply facts from actual experience. Their assistance and cooperation were invaluable.

CONTENTS

MANHATTAN DISTRICT HISTORY
NONSCIENTIFIC ASPECTS OF LOS ALAMOS PROJECT Y
1942 THROUGH 1946

ABSTRACT

Recent wide interest by historians about this era has led the Los Alamos Scientific Laboratory to issue this edited version of an unpublished volume covering the construction, organization, and personnel of Project Y. An Appendix covering some early experiences at S Site has been added.

I. SITE SELECTION

Purpose

By October of 1942, it had become increasingly evident, from the progress of experimental developments supervised by the Manhattan Engineer District at its other installations, that the immediate establishment of an additional research site was necessary for solution of specific problems in production of a nuclear weapon. The purpose of this new installation was the development, final processing, assembly, and testing of the atomic bomb. The contemplated scope of this part of the Manhattan Project was great enough to justify a separate title; accordingly, it was named Project Y.

Requirements

Because the new Project was to be the most secret of the entire Manhattan Engineer District program, isolation was perhaps the first requisite for the site. However, many other factors had to be considered:

The area had to be large enough to provide an adequate testing ground.

The climate had to be such that outdoor work could continue through the winter months.

Access by roads and railroad was necessary for moving in personnel and material.

Sources of construction materials had to be near enough to keep costs reasonably low.

The population within a 100-mile radius of the site had to be sparse, to maintain safety and security.

Utility facilities, including power, water, and fuel supply had to be available or conveniently developable.

Housing facilities had to be present to quarter at least the first personnel.

The ownership and estimated value of land and speed of acquisition had to be considered.

Soil characteristics, timber density, and type of terrain also had to be carefully investigated as the basis for future construction.

The location had to be remote from all sea coasts, as the possibility of attack still had to be considered.

The U.S. Engineer Office and Real Estate Sub-office in Albuquerque surveyed several areas in

New Mexico for this site using these requirements as guidance. Their reports show the emphasis placed upon all of these points.

Sites Considered

Sites at Gallup, Las Vegas, La Ventana, Jemez Springs, and Otowi, New Mexico, were surveyed. After careful investigation, the first three locations were rejected as failing to satisfy the requirements established. Then more detailed reports were made for Jemez Springs and Otowi.

The Southwestern Division Real Estate Branch made a preliminary report on the possibility of locating the site at Jemez Springs, in November 1942. All pertinent factors, such as water supply, housing facilities, access by road and railroad, ownership, and estimated value were considered. A further report, by the U.S. Engineer Office, Albuquerque, New Mexico, covered in more detail the buildings around Jemez Springs which could be used for housing, and considered the sources and costs of construction materials, climate, labor supply, recreational facilities, population within a 100-mile radius, fuel supply, medical facilities, and the steps necessary to acquire the land for the proposed site. Had Jemez Springs been selected, 70% of the housing for the immediate needs of the Project would have had to be built. This report included no recommendations, because the specific purpose of the site was unknown to the office making the survey.

In November 1942, the Manhattan District authorized the Albuquerque Engineer District to conduct a site investigation in the vicinity of the Los Alamos Ranch School, Otowi, New Mexico. Reports comparable to those submitted on the proposed Jemez Springs site were prepared by the Southwestern Division Real Estate Branch and the U.S. Engineer Office, Albuquerque District. The fact that the existing Los Alamos Ranch School buildings could be used for immediate housing was a factor in the recommendation of the site. Further, Otowi was more accessible, had a better water supply and lower valuation, and lay in a more sparsely populated area than Jemez Springs. All of these advantages plus the following favorable points were summed up in the recommendations.

Most of the area (some 46,667 acres of the estimated 54,000 required) could be obtained easily because it was government owned.

The private land was used mainly for grazing, so the purchase cost would be small.

Enough area was available to ensure safe spacing of the Project units.

The nearest town was some 16 miles away, which tended to isolate the site.

The area was located on a mesa, making the entrance thereto easy to control.

The main site was relatively free from timber, and would necessitate little clearing.

Representatives of the Manhattan District, Albuquerque District Office, and the Southwestern Division Real Estate Branch met in November 1942 at the Los Alamos Ranch School to consider that location in detail. The choice of the site was also discussed with Dr. J. R. Oppenheimer, Project Director, and members of his staff, for further confirmation of its desirability. After careful consideration of all the cumulative reports, Major General L. R. Groves determined that Project Y would be centered at the site of the Los Alamos Ranch School, Otowi, New Mexico.

After selection of the site, Lt. Col. J. M. Harman was designated Commanding Officer. The University of California had been selected previously as Operating Contractor to administer the technical work.

Site Description

Los Alamos was in a sparsely populated rural area on the east slope of the Jemez Mountains (part of the Rocky Mountain System), in Sandoval County in north central New Mexico. Approximately two-thirds of the reservation occupied relatively flat, east-sloping bench land, 6900 to 8200 ft above mean sea level, and lying between the valley of the Rio Grande and the east slope of the Jemez Mountains. The western third included the rising east slopes of the Jemez Mountains, 8000 to 9000 ft above mean sea level. The entire area was frequently dissected by east-flowing drainage, and the streams had cut innumerable canyons, 100 to 500 ft deep, separated by mesas of varying extent. The canyon bottoms were generally narrow and rimmed by precipitous cliffs, 50 to 200 ft high. Many of the canyons were box canyons with no access to the mesa on either side.

The project site was about 20 miles, air-line distance, northwest of Santa Fe, the state capital. Santa Fe was the nearest railhead for the project and the terminus of a branch line of the A. T. & S. F. Railway which joined the main line at Lamy, 18 miles farther south.

Access Roads. Access to the Project was over two alternate routes extending westward from paved, primary highway No. 285 that ran northwestward from Santa Fe through Espanola, New Mexico. The shorter route was State Highway No. 4 which left the primary highway at Pojoaque and crossed the Rio Grande River at the Otowi Bridge. This was a secondary, unsurfaced road along an unimproved alignment from Pojoaque to the junction of State Roads No. 4 and No. 5, approximately 1 mile west of Otowi Bridge. This part of the road was 16 to 20 ft wide. In some places the right of way was limited by community buildings located along the highway. Unusually heavy or long vehicles could not use this road because of the Otowi Bridge (Fig. 1.), a single-lane suspension bridge of 250-ft span, 10 ft wide, and designed for 10-ton loading, with its east approach in a narrow side-hill cut requiring an approximately 90° turn on to the bridge. Also the route was so rough and curved as to damage vehicles if frequently traveled. The distance from the Plaza in Santa Fe to the Technical Area at Los Alamos via this shorter route was about 35 miles. During heavy rains this road was closed to traffic by two unstabilized stream crossings, so a longer alternate route via State Road No. 5 and Espanola had to be used.

State Road No. 5 had considerably better alignment and surface than State Road No. 4, but its traffic was also subject to interruptions by high water because of several dips that passed flood waters across the highway. Such interruptions were not so frequent nor so long as on the shorter route. It was 45 miles from the Santa Fe Plaza to the Los Alamos Technical Area via this longer route.

From the junction of State Roads 4 and 5 to the Project, the road was originally rough and poorly aligned. The New Mexico State Highway Department, under the supervision of the Public Roads Administration, built a first-class highway over this part of the access road. The construction was performed under contract by Lowdermilk Brothers Construction Company of Denver, Colorado. Construction (Figs. 2-5) was begun early in September 1943 and completed in December. The work consisted of grading, building drainage structures, and applying base-course surfacing that was wetted and compacted with a roller. During the summer of 1944, the State Highway Department surfaced the entrance highway with a 2-in. bituminous mat and a double asphalt surface treatment.

There was a third access road from the west gate of the Project via Valle Grande to Albuquerque (approximately 110 miles). This route was unimproved, very rough, and closed by snow in winter, so it was not often used.

Structures. Two properties, the Los Alamos Ranch School and Anchor Ranch, had structures usable for housing and storage. The School comprised 54 buildings: 27 houses, dormitories, and living quarters totaling 46,626 ft^2, and 27 miscellaneous buildings: a public school (operated by the state for the children of Ranch School employees), an arts and crafts building, a carpenter shop, a small saw mill, an implement barn, a saddle barn, hog barns, stables, sheds, garages, and an ice house, totaling approximately 29,560 ft^2. Some of these buildings were remodeled into housing and stores and others were entirely removed. There were 4 houses, with approximately 20 rooms, and a small barn at Anchor Ranch Site.

Nomenclature. Because the name "Los Alamos" was considered classified information, the installation was variously identified by Manhattan Project workers as Site "Y," Project "Y," the Zia Project, Santa Fe, Area "L," Shangri La, Happy Valley, and the like. However, the residents of Los Alamos and Santa Fe simply referred to "The Hill" when discussing Los Alamos.

Fig. 1. The bridge at Otowi.

Fig. 2. View west on State Highway No. 4, showing early road construction. Old U. S. Engineer Office photograph.

Fig. 3. View east on State Highway No. 4, showing switchbacks that were eliminated.

Fig. 4. Construction work on Highway No. 4. View south.

Fig. 5. Looking west on the mountain access road with various switchbacks visible.

II. LAND ACQUISITION

Procedures.

In November 1942, the Undersecretary of War directed that the land recommended by the Chief of Engineers to the Commanding General, Services of Supply, be acquired. The necessary procedures were then instituted for acquiring this land for immediate use, the method depending on ownership.

In March 1943, the Secretary of War requested that the Secretary of Agriculture grant authority for the War Department to occupy and use, for as long as the military necessity existed, 45,100 acres of federally owned lands under the jurisdiction of the Forest Service. (See letter, next page.) This authority was granted in April 1943. Arrangements were to be made between the Commanding Officer of the Los Alamos Project and the Regional Forester at Albuquerque, New Mexico, for the prevention and suppression of fires, the management and protection of the area, and the marking of areas within which outsiders might be permitted. It was necessary to withdraw grazing permits authorized by the Forest Service. This was accomplished by direct negotiation between the Real Estate Suboffice and the grazing-permit operators, whereby payment was made on the basis of $20 to $30 for each head of grazing stock.

The process prescribed for acquiring privately owned land was by condemnation or purchase. Authority for condemnation of private lands was contained in the 2nd War Powers Act. Under this Act, the government filed a Petition in Condemnation which resulted in an Order of Possession served by the court on the land owner, who then had to vacate. To acquire the land permanently, a Declaration of Taking was filed by the government and appraisals were made by an appointed commission. If the appraisal was not approved by both the land owner and the government, the case was settled in the U.S. District Court.

History of Acquisition

The land was acquired in five separate areas. In acquiring these areas, it was necessary to acquire all the tracts in fee simple by filing Declaration of Taking proceedings. This was necessary because there was not enough time to negotiate with each owner, and, because condemnation proceedings would still have been necessary to eliminate the numerous title defects that existed. The grazing permits were secured by direct negotiation.

Petitions in Condemnation were filed for the various areas between May and December 1943. Priority and rate of acquiring the areas were determined by the Commanding Officer through the Southwestern Division Real Estate Suboffice. The Petitions were worded to cover all grazing, mineral, special-use, water, and timber permits, and all other interests of whatever nature, so that private individuals would have no reason whatsoever to enter the areas.

The land was held as follows: 45,737 acres by permit from other government agencies, 3600 acres by War Department ownership, 40 acres by lease (returned to owner October 17, 1943), and 6 acres by easement. This included the right of way for a power line to the Project from the New Mexico Power Company's Bernalillo-Santa Fe line.

The Los Alamos Ranch School was appraised by Major Gerald T. Hart, Division Engineer Office, Dallas, Texas, and Watson Bowes of Denver, Colorado, a member of the American Institute of Real Estate Appraisers. All other appraisals were made through direction of the Real Estate Suboffice, Albuquerque, New Mexico. All factors affecting market value were considered, including desirability, physical characteristics, location, soil type, and topography. Prices at which similar land was selling in the immediate area were the main criteria used in establishing values.

For the 22 tracts in the first four areas that were condemned and for which Declaration of Taking proceedings were filed, 16 stipulations were secured to acquire title in fee simple, indicating that the prices offered were satisfactory. The principal objectors to the prices offered were Elfego Gomez of Espanola and Manuel Lujan, of Santa Fe. Of the 13 privately owned tracts in the fifth area, 6 were acquired by direct negotiations and 7 were condemned under the Declaration of Taking Act because of price controversy or title defects. The principal objectors to the amount

CE 601.52 SPELD
Los Alamos, N.M. 22 March 1943

The Honorable,

The Secretary of Agriculture.

Dear Mr. Secretary:

There is a military necessity for the acquisition of approximately 54,000 acres of land for the establishment of a demolition range.

The required area is located near Santa Fe, and within Sandoval County, New Mexico, as shown outlined on the enclosed map. Title to 8,900 acres of privately-owned land in the area will be acquired by condemnation or purchase. The remaining 45,100 acres of federally-owned lands are under the jurisdiction of the Forest Service.

It is requested that you grant the War Department permission to occupy and use all of the federally-owned lands within the above-described area for so long as the military necessity continues.

Your cooperation in making these lands available to the War Department will be greatly appreciated.

Sincerely yours,

s/HENRY L. STIMSON
Secretary of War

offered were Elfego Gomez, and Ernesto and Adolfo Montoyo of Espanola.

All Condemnations and legal matters regarding government possession of the private lands and satisfaction of claims against the public land were handled by the Southwestern Division Real Estate Suboffice, in Albuquerque. Other than those usual in condemnation proceedings, no special problems were encountered. However, part of the Project was bordered by the Bandelier National Monument and by an Indian sacred burial ground that caused an irregularity in the southeast boundary.

The right of entry to the Los Alamos Ranch School granted by Mr. A. J. Connell, President and Director, in November 1942, to construct, survey, and explore on the School lands and property, greatly facilitated the early beginning of construction. This property was then acquired by direct negotiation. All other areas were acquired when needed by the Project on the request of the Commanding Officer to the Real Estate Suboffice. Because the right to use and occupy the land was acquired quickly, the Project was not hindered, even though transfer of title required the usual processing time.

Exclusive federal jurisdiction over all land in New Mexico acquired by the United States for military purposes was accepted by the Secretary of War in August 1944 and May 1945, in accordance with New Mexico Statutes. These lands included a material portion of Los Alamos Reservation, particularly the land acquired from the Los Alamos Ranch School, Anchor Ranch, and certain individuals. However, because the laws of the State of New Mexico did not cede exclusive jurisdiction over land reserved from the public domain for military purposes, there was some question about the extent of the U.S. Government's exclusive jurisdiction over the Los Alamos Reservation.

The original estimate of land required was 54,000 acres; this was later decreased to 49,383 acres. Some 45,667 acres were obtained from the Secretary of Agriculture, the rest being acquired for the power-line right of way as follows: 46.5 acres from the Department of Interior, Grazing Service; 3.5 acres from the Department of Agriculture, Soil Conservation Service; and 19.5 acres from the Department of Interior, Indian Service. The War Department acquired ownership of 3,600 acres. Forty acres surrounding and including Frijoles Lodge were leased in June 1943 for use as additional housing. This lease was terminated in October 1943 but used again in July and August 1944 during the peak housing shortage. The power-line right of way also crossed private land and New Mexico state land, so easements were secured for 1.242 acres in the former case and 5.190 acres in the latter.

Costs

The total cost of all lands acquired in fee simple, by easements, by lease, and in any other manner was $414,971.00. The cost of the school, with buildings, appurtenances, and 470 acres was $350,000.00; Anchor Ranch, 322.16 acres, was another high contributor to cost at $25,000.00.

III. DESIGN AND ENGINEERING

Design and engineering, through March 1944, were supervised by the Albuquerque Engineer District of the Southwestern Engineer Division. Then the Manhattan Engineer District assumed supervision. Because technical design and engineering depended largely upon the development of research and experimental processes, actual needs and requirements continually demanded major or minor alterations. The firm of W. C. Kruger was selected as architect-engineer for initial design and engineering and continued its work under Manhattan District supervision. R. O. Ruble was the Engineer, and R. W. Graef was the Supervisory Architect for the firm. In December 1945, Black and Veatch were also engaged as Consulting Engineers for all utility designing and engineering.

The Albuquerque District negotiated a contract for design of the originally authorized buildings and utilities with the Kruger Company because they maintained a competent architectural and engineering staff and had done considerable satisfactory work for the District. Further, their office was in Santa Fe, in a good position to collaborate with the Operating Contractor about special technical items not ordinarily covered in standard Army construction. The original contract was for providing plans to adapt existing buildings and plans for new buildings. The original directive covered rehabilitation of 31 buildings and the drawing of plans or adaptation of Standard War Department construction plans for 111 buildings, together with planning of utilities and streets. This new construction was to provide housing and other facilities for military personnel, apartment houses and dormitories for technical personnel, a power plant, administration buildings, dispensary, school, Post Exchange, commissary, theaters, and technical buildings. As the Project expanded, the work required of Kruger increased in proportion.

The normal method of determining requirements for technical-area structures, was for the University of California to sketch the necessary floor plan and list utilities and other major fixtures; then to submit these data to the local Engineering Division, which contacted the construction company and outlined the job. The completed drawings were checked for accuracy and standard-practice engineering, altered if necessary, and submitted by the Engineering Division to the University of California for final approval. For nontechnical structures, the requirements were transmitted to the Architect-Engineer directly by the Project. The Architect-Engineer's representatives worked directly with the originator of the requirements, if details were incomplete or inaccurate. Project Headquarters maintained a small engineering organization for design and engineering of small, simple tasks for which outside contracts were not considered justified or desirable.

All design and engineering contracts were awarded to the one firm, W. C. Kruger Company, Architect-Engineers, to minimize the delay that normally accompanies emergency work. Often actual floor-plan construction was in process before completion of superstructure drawings. The process by which contracts were awarded or extended was to outline the required work to the firm, receive their cost estimate, compare it with costs of similar jobs, completed or in progress, and then negotiate with the University of California in case of differences. Some contracts were renegotiated when it appeared that excessive profit had been made. Profit was figured by the local Engineer Division using the Corps of Engineer standards, with payroll costs, overhead, current material costs, etc. as a basis of accounting. Whenever excessive profits were made, the contractor returned the established overage to the Government.

Fees for the architect-engineering work were determined through negotiations with the chosen architect, and on the first job, Kruger worked his men for the Government on an hourly basis. Additional contracts were negotiated under which the firm would prepare plans for new construction, provide survey and layout crews, and do a small amount of inspection to coordinate with that done by the Engineering Division concerned. A lump-sum price was negotiated for each item of work, and a check revealed that the cost of Architect-Engineering services never exceeded 3½% of the cost of construction, the average being 2.8 to 2.9%. Kruger was paid $174,968.98 by the Albuquerque District, and $568,737.70 by the Manhattan District, totaling $743,706.68, through December 1946. Black and Veatch were paid $164,116.00 through December 1946.

IV. CONSTRUCTION

General

All construction through mid-March 1944 (except that covered by Contract W-7425-ENG-15, handled by the Manhattan District) was supervised by the Albuquerque Engineer District. Then supervision was transferred to the Manhattan Engineer District (Project Y Headquarters). Also, during the early period of original construction, in an effort to expedite the work on some of the new buildings, the University of California hired construction workers independently, under their own supervision.

Construction accomplished under the Albuquerque District cost $9,300,000.00; that under the Manhattan Engineer District, including utilities and hired labor cost about $30,400,000. These figures do not include architect-engineer services. Contracts were usually negotiated lump-sum contracts or purchase contracts at a stipulated lump sum. They were awarded only to reputable firms, either with or without competition. As mentioned, speed was the most essential factor, and contracts were awarded to firms considered most capable of carrying out the terms of their agreements. Construction progressed as well as could be expected given the various unusual difficulties peculiar to the Project.

Selection of Contractors

Responsibility for the original selection of the construction contractor was left primarily to the District Engineer, Albuquerque Engineer District. This selection was subject to approval by the Manhattan District, so that security provisions might be taken into account. To determine which construction contractor would be best for this particular job, the Albuquerque District considered all contractors in this region who had enough capital, personnel, and equipment to accomplish the work within the time required. In the process of elimination, the District decided that the M. M. Sundt Construction Company would be available to do the work in the time required and was suitable for this type of work. Negotiations were begun, based upon a Government estimate, and, in December 1942, Sundt was awarded the contract. This company owned and operated a large number of trucks, which made them desirable, because materials had to be hauled from Santa Fe to the job site. They did not use subcontractors, but had their own plumbing, electrical, and painting departments, and, as a result, were the major contractor on the Project during initial construction. This type of organization was desirable from a security standpoint as it simplified control over its personnel.

Other contractors to do road work, erect tanks, etc. were selected using the same primary criteria. A. O. Peabody, of Santa Fe, was awarded the contract for surfacing roads and streets within the Project. The contract was negotiated using a Government estimate based on past experience of the Albuquerque District, and called for the construction of streets and roads within the Project area, including construction of eight miles of road to the area then known as the Anchor Ranch Range. Roads and streets were of extreme importance to the Project because of the dangerous and delicate materials to be transported. Therefore, every effort was made to construct satisfactory all-weather roads. Construction and surfacing of the main road in the early part of the program contributed to a large degree in lowering the cost of constructing the Project, by facilitating transportation of men and materials. Lowdermilk Brothers, Denver, Colorado, was awarded a contract to furnish gravel for surfacing site roads and Post parking areas.

In October 1943, a lump-sum contract was negotiated with the Perdue Tank Company of Wichita, Kansas, for construction of two 150,000-gal redwood water-storage tanks (Fig. 6). The contract was later modified to include re-jointing and resetting one 300,000-gal redwood tank. Sundt Construction Company built a 250,000-gal elevated wooden storage tank (Fig. 7) to provide required water pressure through the distribution system. A million-gal steel tank (Fig. 8) replaced this tank late in 1946.

The original technical buildings constructed by Sundt were J, M, S, T, U, V, W, X, Y, Z (Figs. 9-19), and Boiler House No. 2. Building T was the original University of California headquarters, containing substantially all phases of its work except the laboratory functions and the shop and

Fig. 6. Two 150,000-gal redwood water-storage tanks. The truck in right center was one used to transport water from the Rio Grande Valley. A pumping station to remove water from the trucks is at its right. In the background is S Building, a Technical Area storeroom. Part of the Technical Area fence is shown at the right.

Fig. 7. The 250,000-gal tank. The building in the left center is an original shop of the Los Alamos Ranch School which was converted into the Housing and Express Office.

Fig. 8. The million-gallon steel water tank.

Fig. 9. Technical Warehouse, S Building.

Fig. 10. Original Administration Building, T-1.

Fig. 11. A later Administration Building, T-44. It provided offices for the Commanding Officer, Executive Officer, Chief Clerk, and Mail and Records Section.

Fig. 12. Technical Building U. Another of the original Technical Area buildings, used for electronics.

Fig. 13. Technical Building V. One of the first technical buildings, used for shops.

Fig. 14. Technical Building W. Another of the original Tech Area buildings, it housed the Van de Graaff accelerator.

Fig. 15. Technical Building X, constructed early in 1943. It housed the cyclotron.

Fig. 16. Technical Building Y. One of the first Technical buildings, it was a physics laboratory.

Fig. 17. Technical Building Z, built early in 1943. It contained the Cockroft-Walton accelerator.

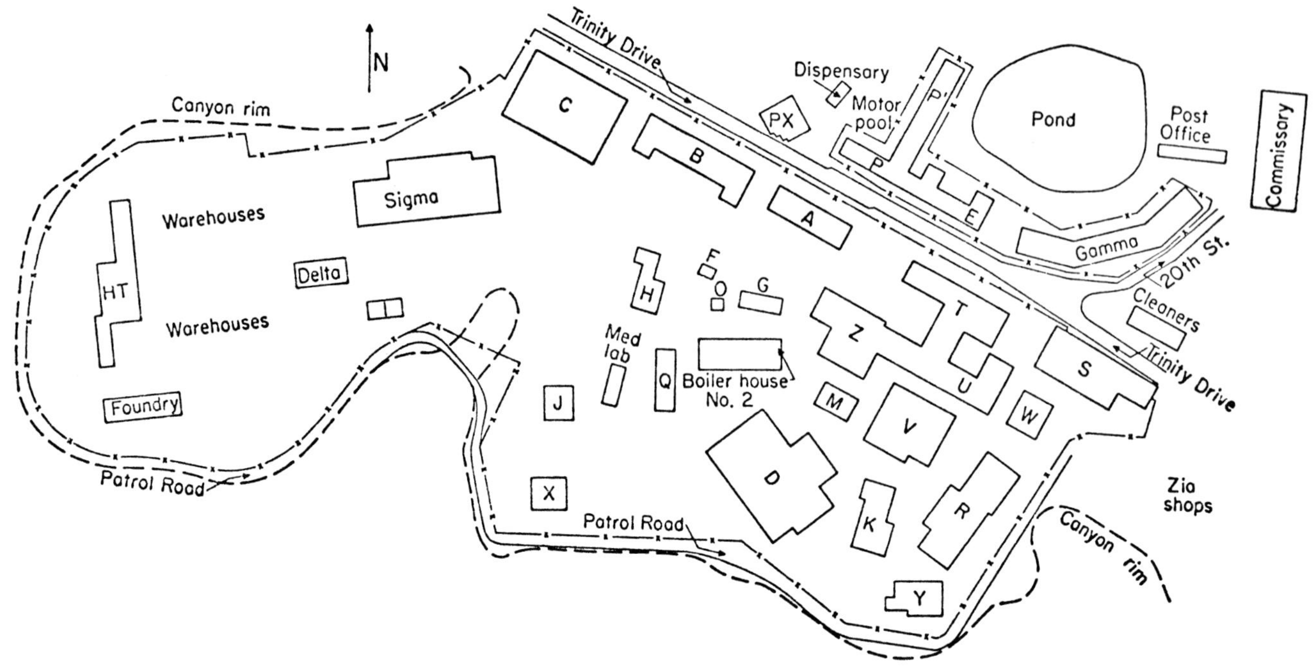

Fig. 18. Technical Area map.

Fig. 19. Gamma Building.

storeroom. The laboratory equipment, stock, and shops were housed in the other buildings. Later, Sundt added buildings A, B, and C (Fig. 20). Building A then became the Administration Headquarters; Building B provided additional laboratory, office, and conference space; and Building C was another shop. Sundt also constructed buildings D and Q (Figs. 21 and 22). Force Account constructed buildings G and O in 1943, as well as modifying building K, an original Ranch School building.

After Sundt completed their contract in November 1943 and moved their personnel and equipment from the Project, it became evident that more work would be required, enough to demand a construction contractor. The University of California had outlined preliminary requirements for S Site. Robert E. McKee, General Contractor, El Paso, Texas, was engaged to perform the work at S Site under negotiated lump-sum contracts. This firm was one of the largest in this part of the country, and had constructed many important projects since the beginning of the war. It was selected for this reason and because of its strong financial position, adequate equipment, and personnel. MeKee had subcontractors for the electrical, plumbing, painting, roofing, insulating, fire-control sprinkler, and spark-proof-flooring installations.

During construction of S Site, the University of California asked for a 30- by 100-ft, two-story building (P Building), and an overhead covered passage to connect it with the overhead covered passage between buildings A and T, originally built by Sundt (Figs. 23-25). This work was done under the supervision of the Albuquerque District. Competitive bids were taken from J. E. Morgan and Sons, Contractors, and from McKee, both of El Paso, Texas, and were opened by the Commanding Officer, Project Y, on March 1, 1944. McKee was low bidder and was awarded the contract.

A directive by the Albuquerque District in January 1944 authorized construction of housing for 56 families, to consist of 8 prefabricated one-bedrood duplexes, 15 prefabricated two-bedroom duplexes, and 5 prefabricated three-bedroom duplexes, together with the necessary utilities, including water, sewer, and electrical distribution systems, and the necessary road construction. This work, supervised by the Albuquerque District, with Architect-Engineer service by Kruger of Santa Fe, used a schematic floor plan furnished by the University of California.

Bids were received from three prefabricating manufacturers, and it was decided that the unit furnished by the Houston Ready-Cut House Company, Houston, Texas, most nearly met the requirements. Morgan and Sons, being a reputable contractor and having performed previous satisfactory work, was selected to erect the buildings and construct utilities. The negotiated lump-sum contract was awarded on a basis of unit prices determined from previous similar construction supervised by the Albuquerque District.

Upon completion of the housing units by Morgan and of P Building and S Site by McKee, the Albuquerque District was relieved, in March 1944, of further construction supervision and was replaced by the Manhattan District. This was done because it was thought that most of the construction was finished and that to meet the Operating Contractor's need for considerable remodeling, small additions, and shop work, it would be necessary to maintain only a number of Project construction crews (Force Account). The plan was to have these crews do all construction work at the Project.

Because the whole program was not outlined at once, and its continued growth was not anticipated, it was felt that these crews could accomplish all additional construction required. However, shortly after the completion of S Site, it became evident that the work being planned was much more than could be accomplished without a construction contractor for the large tasks. Construction planned after the Manhattan District assumed supervision was awarded to Robert E. McKee by negotiated lump-sum contracts, using unit prices determined from past contract records and from comparison with Project construction-crew costs.

Because of the haste with which much of the construction had to be completed, work sometimes started before completed plans for the entire development were available; therefore, part of the work was started before a price for the entire amount was agreed upon. The price negotiated for work already started was considered fair by both parties. In fact, the contractor never tried to obtain excessive prices for work performed.

In a number of cases, buildings substantially like those built by the Sundt Construction Company under the original contract were

Fig. 20. Technical Building B, (rear view) looking northeast.

Fig. 21. Technical Building D, looking northeast.

Fig. 22. Technical Building Q, Headquarters for the Health Group.

Fig. 23. West elevation of two covered passages over main road. P Building is on the left. The Technical Area Pass Office is on the right side of the road. Building T is on the right, partly hidden by the bridge.

Fig 24. Technical Building A, connected by passageway to Building B, which was very similar to A in design and construction.

Fig. 25. Passageway from A Building to B Building on the left, looking north.

required; then the price of the original contract and McKee's price were compared. Usually, McKee's prices were found to be lower. This was attributed partly to the fact that the access road was better during the latter construction; also, labor was easier to obtain.

Force Account

It was necessary to establish a construction and maintenance organization that could maintain existing facilities and make many necessary additions and alterations. This was done by direct government employees, or Force Account. Competent foremen of the various trades, with crews to work under their supervision, were hired by the Project. It was eventually necessary to hire a force larger than that usually required in Army installations, because it was soon found that this organization would have to handle a great deal of new construction. Normal policy would have been to contract all this new construction. However, because many of the items were small and did not require plans and specifications, they were undertaken by Force Account crews.

Speed was the most essential factor in all the construction undertaken. Work was required not only in the main Technical Area but also in outlying sites. To do this work with minimal interruption of the University of California's activity, it was necessary to use construction forces that had been cleared by the Project Security Division. Therefore, Force Account undertook many jobs that would normally have been handled by a construction contractor. It was felt that it would be better to have small Force Account crews working continually in certain areas than to have contractor's men moving in and out at intervals, perhaps with different men in the crews each time, thereby increasing the possibility of publicity about the contents of the buildings. Throughout the entire Project much of the work had to be done at night to prevent interrupting the operations of the Operating Contractor. Often, construction and maintenance crews working at outlying sites or in the Technical Area had to be removed because of highly secret or dangerous University of California operations. This caused a great loss of time.

Force Account crews maintained a utility yard near the Anchor Ranch Range, and also provided personnel for maintenance shops operated by the University of California in the main Technical Area itself. Force Account work was divided into Post Area, Housing Area, Special Assignments, Technical Area, and outlying areas. The actual work was divided into new construction, maintenance, and operation.

The original plan was to use military personnel, with only a minimum of civilians, to keep housing requirements minimal and to provide better security. As the Project grew and as military personnel were urgently needed in combat areas, it became more and more difficult to obtain skilled tradesmen from the ranks of enlisted personnel. Therefore, it was necessary to supplement the organization by hiring civilians. Skilled civilian tradesmen and skilled enlisted personnel frequently worked closely together. Sometimes civilian foremen were in charge of enlisted crews, and at other times the situation was reversed.

There was a constant problem of maintaining harmonious relations between enlisted personnel and civilians because of the pay differential. Usually, however, difficulties resulted from a clash of individual personalities. As the Project progressed and adequate civilian personnel became available, it became possible to separate the civilian and military crews to a larger extent. After March 1945, the tendency was to release Army personnel and replace them with Civil Service employees. The Force Account was reduced when the construction work was again assigned to outside contractors, in particular McKee. It was further reduced when the Zia Company was established in April 1946.

Description of Work

Buildings and Equipment. Buildings at the Project were of three general classifications: housing for military and civilian personnel, technical buildings, and administration and utilities buildings. Buildings for military personnel were of the regular Theatre of Operations type. The original housing for civilians was mostly of economical semipermanent type. Some of the apartment buildings had wood siding, but most had triple-seal gypsum board siding. Interiors were sealed with gypsum board or Celotex. Plans for a permanent housing area drawn by W. C. Kruger Company were approved in June 1946. These buildings were of stucco and concrete brick. Administration and utility buildings (including schools, hospital, etc.) had wood siding with interiors of ½-in. gypsum board.

The original technical buildings were of modified mobilization type, with exteriors of drop siding on gypsum-board sheathing and pitched roofs covered with asphalt shingles. Other buildings were modified mobilization type with cement asbestos-shingle exteriors and interior gypsum-board sheathing.

Some of the Technical Area laboratory buildings had complete air conditioning and dustproof construction. Some had ceilings of acoustical tile, others were constructed of a single layer of 1-in., triple-seal, gypsum-board siding nailed directly to the exterior of the studs and a single layer nailed to the interior. All exterior joints were sealed with mastic. Most of the buildings in this group had automatic sprinkler systems for fire protection. Generally, technical buildings were heated from a central boiler plant. The Technical Area also had an extensive public-address system.

Outlying Satellite Sites. A total of 25 sites in addition to the Main Technical Area were built for experimental purposes (Figs. 26-37). Because many of these sites were located in rough unimproved terrain (Fig. 38), roads, utilities, and building construction were necessary before they could be used.

Utilities. ***Roads, Parking Areas, and Walks.*** Roads on the Project, constructed by A. O. Peabody, by Sundt, and by Force Account crews, included 22 miles of dirt roads, 27 miles of gravel roads, and 17 miles of bituminous-surfaced roads. The main road through the Project to S Site was bituminous surfaced, 20 ft wide, and some 8 miles long. Bituminous-surfaced roads through the housing areas were 16 ft wide. Gravel roads throughout the sites were generally 18 ft wide, and dirt roads were 8 to 20 ft wide.

Many of the access roads had to be built in winter when the ground was frozen. Gravel surfacing material was hauled approximately 15 miles from a pit and screening plant on State Highway No. 4. All roads were designed to follow the natural contour of the ground as much as possible, to minimize cut and fill. Few bridges were constructed, dips or concrete culvert pipe being used to solve the drainage problem.

Fig. 26. East elevation of Building S-1, office building at S Site, showing entrance to equipment room, first aid room, and east end of the building.

Fig. 27. South elevation of Building S-24, a casting building at S Site. Note the wooden barricade at left.

Fig. 28. Interior of work room in Building S-24.

Fig. 29. Wooden and earth-filled barricade near Building S-24, S Site.

Fig. 30. Northeastern elevation of Building S-28, an S-Site shop and laboratory.

Fig. 31. Southeastern elevation of Building S-67, a magazine at S Site.

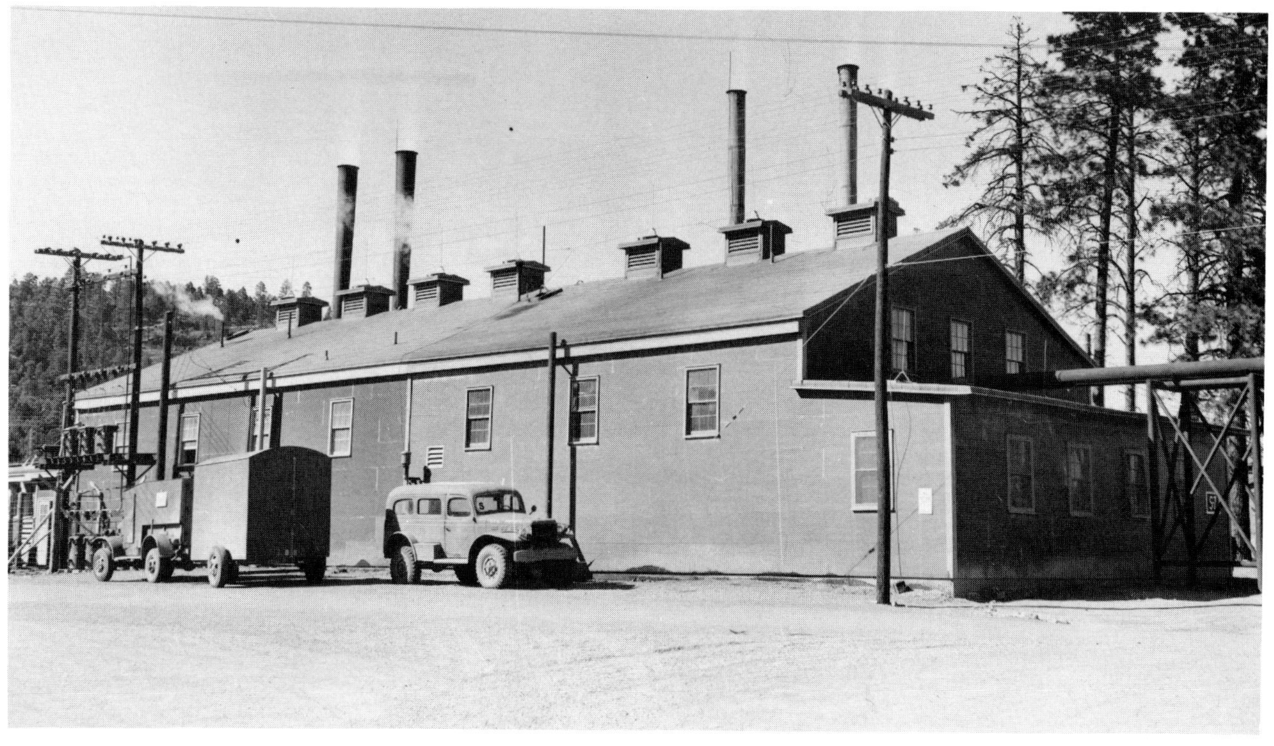

Fig. 32. Southeastern view of Building S-7, the boiler house at S Site.

Fig. 33. Boilers in Building S-7.

Fig. 34. Expansion loop in steam distribution line at S Site.

Fig. 35. S Site in 1949.

Fig. 36. DP Site, the plutonium-processing plant.

Fig. 37. Scene at DP Site.

Fig. 38. Technical Building N. This is an excellent example of the rough terrain on which structures were erected.

Under the original construction, 24,000 square yards of parking areas were constructed, with single bituminous surface treatment on a gravel base. As the job progressed, it became essential to provide 40,450 square yards of additional parking area.

Unstabilized gravel-surface walks were constructed to the front entrances of apartments and dormitories. These walks were 4 ft wide and bordered by rough, untreated 2 by 4 timbers to retain the surfacing. A 5-ft-wide gravel walk, treated with one application of asphalt, was constructed from the Technical Area to the West Mess to alleviate the hazards of pedestrian traffic on the main street. There were other short sections of walk where automobile and pedestrian traffic was heavy.

Electric Power Facilities. During original construction, three used Worthington electric power generators were purchased, with the agreement that they would be reconditioned by the vendor. However, because of the urgency for power, it was necessary to purchase the unconditioned units, and recondition them after installation. With the expansion that started about the time the power plant (Figs. 39, 40) was finished, it became evident that the three units would not provide enough power. As it was impossible to shut down any of the units for necessary overhauling, the power plant was expanded.

Two generating sets, each rated at 1000-kW output, together with all necessary control mechanisms and operating auxiliaries, were added. The sum of the sea-level-rated capacities, corrected to operation at 7300 ft above sea level, was an estimated 2193-kW actual output. The diesel engines were started by air at 250-lb/in.^{2} pressure, supplied by air compressors. These engines operated on grade 2-102 fuel oil with a specific gravity of 38.2. Fuel oil was stored in steel tanks, supported on concrete cradles, approximately 100 ft west of the power plant. All piping was enclosed in a concrete tunnel between the tanks and power-plant building.

Control of the four outgoing primary circuits was obtained at a switching station, a four-pole structure with cross timbers for mounting disconnects and busses. The station was designed so that all feeders from the power plant to the switching station could be paralleled, or any one area could be fed from any of the four feeders from the plant. Later, outdoor cubicles and additional switch gear within the plant were added. These changes, in June 1946, provided 26 outdoor cubicles, with 19 circuit panels inside the plant proper.

Because of further expansion, it became apparent early in 1944 that additional power would be required. Detailed records were kept of the power actually used. In August and September, the load on the plant exceeded the maximum for safe operation over an extended period. To accommodate the peak loads during the day, it was necessary to operate all five units of the plant, and, had one of the larger units been out of order, it would have been necessary to disconnect some of the load from the power plant. Because electric power of standard voltage and frequency was such an essential factor in the Technical Group's work, it was essential to provide additional power immediately.

The solution was approached from two viewpoints. One was to provide an additional diesel-driven generator in the existing power plant. The other was to construct an approximately 25-mile-long high-voltage power line from a point where it would tie into an existing New Mexico Power Company transmission line. Investigation revealed that the estimated cost of constructing the power

Fig. 39. The Electric Power House.

Fig. 40. Electric Power Plant, interior view.

line, to supply between 1000 and 2000 kW, would be $156,500 and that of providing a new generating unit to supply 1000 kW would be $144,228. The line could be built far sooner than the generating unit could be purchased and installed. Because it was felt that the power line would allow greater flexibility, that operating costs were comparable, and that it would possibly allow for future expansion, construction was begun in October 1944, and completed during February 1945. The power line consisted of Copperweld conductors supported by single-pole construction for nominal straight-line runs and wood H-frame construction for turnout angles exceeding 15 degrees. Its maximum capacity was 5000 kW, with a voltage drop of 5%. It was built by the Reynolds Electric Company under a subcontract with R. E. McKee, for $167,457.00, and was operated and maintained by the Government. Because the New Mexico Power Company had limited capacity, there was no firm contract for any definite amount of power. Instead, there was a "dump" contract that was, in reality, a "gentleman's agreement" to supply the Project with as much power, estimated at 1000 to 2000 kW, as was available after the Power Company's regular requirements were fulfilled.

In April 1945, the University of California and Post Area authorities estimated the power needs through October and forecast a peak load of 5300 kW, which exceeded the combined output of the power plant and the available facilities of the New Mexico Power Company. Planned expansion of the Power Company facilities promised only a 1000-kW increase, and even that could not be obtained before January 1946. Two 1000-kW diesel-engine generating units were procured and installed, but their operation was delayed until January 1946, because some of the auxiliary equipment was damaged in shipment.

The electric distribution system was supplied through the switching equipment near the power plant. Primary feed from the plant was 3-phase, 60-cycle, 2400-V, distributed through 12 primary circuits. Three circuits served the Technical Area; one, the troop housing and service area; and the other eight, the rest of the post housing, administration area, and outlying sites. The voltage was stepped up to 13.2 kV for 1 of the 12 circuits that served the S, P, R, Two-Mile Mesa, and South Mesa sites.

Except for a little underground cable at the outlying sites, all power was distributed on pole lines; high-voltage poles being 35 to 45 ft tall, and low-voltage poles 25 to 35 ft tall. The high-voltage lines were slightly less than 8 miles long, and conductors were of bare No. 4 copper wire. The main area was served by 22,755 ft of primary distribution lines, 18,200 ft of secondary lines, and 19,811 ft of service lines. Conductors, consisting of No. 4 to No. 10 copper wire, were carried on some 419 poles. Stepdown transformers were generally mounted on the poles close to power needs.

In late 1946, it became apparent that still further expansion would be needed to provide adequate power. It was impossible to increase the supply from the New Mexico Power Company, so Black and Veatch, Kansas City, Missouri, Consulting Engineers, investigated the situation and recommended that two 1715-kW diesel-engine generating units be procured. Bids were accepted, and these units were purchased from Nordberg Manufacturing Company, Milwaukee, Wisconsin. The addition also called for expanding the power-plant building and panel boards, as well as revamping some of the distributing systems, substations, installations, and transformers.

Water Supply. Water was obtained from five main sources: Los Alamos Canyon, Water Canyon, Pajarito Canyon, Guaje Canyon, and by a pipeline from wells near the Rio Grande River. A small supply came from American Spring south of Water Canyon. These various sources were developed by stages as the scope of the Project and its demand for water increased. Extensive studies of the most economical method of supplying the Project's water needs were made from time to time, and the results were embodied in a "Report of Water Supply, Los Alamos Project, Los Alamos, New Mexico" prepared by the U.S. Engineer Office, Albuquerque, and R. O. Ruble, Consulting Engineer, Santa Fe, in October 1943.

The Los Alamos Ranch School had obtained its water from Los Alamos Creek through a 6-in. pipeline, fed from a small reservoir created by construction of an earth-fill dam across the creek channel. When the property was acquired the dam was incomplete. During the initial Project construction, the dam was heightened, a concrete spillway was built, and riprapping and other incidental construction work were completed. The resulting reservoir was believed adequate for the Project as planned. As the Project grew, however, it became necessary to augment the water supply, and the Water Canyon, American Spring, and Pajarito Canyon sources were developed. Later, further Project growth, extreme drouth, and algal pollution of the water impounded in the Los Alamos reservoir caused another water deficiency. A line laid on the ground surface to Guaje Canyon to alleviate the emergency was regarded merely as a temporary expedient. Then reanalysis of Project water demands and a more complete study of minimum stream flows indicated that it was necessary to maintain a continuous flow from Guaje Canyon. This line was then winterized by burying it, mounding it with earth, or wrapping it with insulation, with the topography determining the cheapest method.

Pipelines included a 6-in. line from the Water Canyon source to its junction with the Los Alamos Canyon line, from which point one 6-in. and one 8-in. line extended directly to the Project. The Pajarito Canyon creek water was transferred from its source by a 6-in. pipeline, (replacing an open flume) to a small earth reservoir near Anchor Ranch, thence through a 6-in. line to the connections at the site, with unused water continuing through the 6-in. line to a connection with the Water Canyon 6-in. line. The Guaje Canyon supply line consisted of 4-in., 4½-in., and 6-in. pipe. Flow from all these sources to the Project was by gravity.

The booster-pump station on the Guaje Canyon line increased the rate of flow from the source to the Project and provided delivery from the source to the Los Alamos storage reservoir. The reservoir was kept as full as possible at all times, all excess flow from pipeline deliveries being sent to it. The small earth reservoir at Anchor Ranch was considered dead storage for fire-control reserve for the Anchor Ranch Site. Ashley Pond, a Ranch School swimming pool, adjacent to the north side of the Technical Area, was also converted to a fire reservoir. The 13.5-million-gal reservoir at Los Alamos Canyon was the main storage reservoir. The 300,000-gal tank on the Water Canyon line, known as the equalizing reservoir, was operating storage. A 250,000-gal elevated wood tank was constructed to provide distribution pressure, and four 150,000-gal wood tanks, were used for storage. Water from the reservoir to these tanks passed through either the 6- or 8-in. line to a chlorination house at the west end of the main area. Distribution to the fire-sprinkler system and fire plugs was direct to the main from the reservoir. Connections to the pressure tank were possible, to ensure protection in case one source was not in operation. Automatic-starting pumps provided pressure to the sprinkler system. Water-distribution mains in the central area included: 3,650 ft of 6-in. pipe, 12,400 ft of 8-in. pipe, 1,750 ft of 10-in. pipe, and 250 ft of 12-in. pipe. This system was augmented in December 1945 and January and February 1946 by delivery of approximately 300,000 gal a day by truck from the Rio Grande River Valley, approximately 20 miles away. This was necessitated by mechanical breakdown and freezing of one pipeline, and by an unusually dry season with excessive demand.

In August 1946, three wells were drilled in the Rio Grande Valley. Approximately nine miles of 14-in. pipe connected these wells to four pumping stations, which in turn lifted the water 1800 ft vertically to a new 1,000,000-gal standard steel storage tank. This tank was completed in November 1946 and replaced the 250,000-gal elevated wooden tank. As of December 1946, the permanent pumps were not installed, but the new water supply was being used by means of temporary gasoline-driven pumps. A new power line to feed these pumps was to be completed by August 1947, at which time the permanent pumps were also to be in place.

Technical-Area Steam Plant. A steam plant, called Boiler House No. 1 (Fig. 41), was built to supply steam to the first technical buildings constructed. Two second-hand, coal-burning, hand-fired, 100-hp locomotive-type boilers, together with pertinent equipment, were installed. A network of concrete tunnels from the boiler house to the various buildings enclosed the steam lines, pump-condensate return piping, and compressed-air piping.

A second steam plant (Fig. 42) was soon constructed to serve additional technical buildings. Boilers in this plant were two Kewanee No. 586 and three No. 587 fire-tube boilers, all of which were coal burning and fired by Riley Jones hydraulic ram-type stokers. The five boilers' total capacity was 826 hp. In addition to the regular automatic water-feed regulators and automatic control of the feed-water heater, there was complete automatic regulation of the combustion rate. After one year of operation, the two original boilers in plant No. 1 were leaking so badly that

Fig. 41. Boiler House No. 1, later modified into Technical Building M. V Building is in right background.

Fig. 42. Boiler House No. 2.

they would have to be replaced if that plant were continued in use. So, Boiler House No. 1 was dismantled, and a 174-hp boiler was added to Boiler House No. 2. Later, these six boilers were converted from coal stokers to oil firing.

Sewage System. The sewage system soon became too small because the progressive expansions of the Project could not be foreseen during the original planning. A factor that led to the original sewage-disposal system, numerous septic tanks serving small areas, was the irregular terrain, which made it impracticable to have all sewage collected in a central sewage-disposal system without installing pumps. At the beginning, when a central system was considered, the small number of people served made it more economical to install the septic-tank system. The additions to the Project overloaded the tanks, and, although new ones were constructed, it proved more economical to clean the tanks oftener than to construct more of them. The immature effluent from these tanks was drained to nearby canyon floors, where it soon disappeared into the ground. In 1946, the system in the central area included: 61 manholes; 30,035 ft of 8-in. vitrified sewage pipe and 7,265 ft of similar 6-in. pipe for distribution lines; 3,796 ft of similar 8-in. pipe and 23,091 ft of similar 6-in. or smaller pipe for service lines; and approximately eight concrete septic tanks. There were also numerous small tanks and systems at the outlying sites.

Robert McKee was awarded a contract for a new central sewage-disposal plant to replace the overloaded and obnoxious septic-tank system. This was to consist of collecting lines, booster pumping stations, and a modern trickling-filter disposal plant, to be in service by September 1947, at which time all the septic tanks in the main Project area were to be demolished, thoroughly sanitized, and filled in. Use of the effluent from the sewage-disposal plant for irrigation to beautify various areas of the Project was being investigated.

Telephone System. When the land for the Project was acquired, telephone communication from the area to the Mountain States Telephone and Telegraph Company lines was by a Forest Service line of No. 9 galvanized iron wire on native untreated poles. This was soon replaced by an Army field-wire connection to the same system, but direct to the Ranch School site. In February 1943, the telephone company was asked to construct two 18-mile-long connections from their lines at Pojoaque, and these lines were finished six days later.

The PBX board was originally installed in the Administration Building, with a subboard in T Building (Fig. 43), and tie lines between the two boards. A year later the two boards were combined in T Building. Six months later the installation was moved to A Building, in the Technical Area and enlarged to a three-position board. It was later increased to a six-position board. All work was done by Mountain States Telephone and Telegraph Company under contract. In March 1945, the total number of lines to the Project was raised from five to eight. Six additional lines and a complete dial system, to permit installation of private telephones in residences, which theretofore had not been permitted, were to be installed by July 1947.

Teletype. The first teletype machine was installed in the Bishop Building in Santa Fe in March 1943. This was moved to the Project and installed in T Building by the Mountain States Telephone and Telegraph Company. The Army Engineers installed a coding machine on this TWX machine. In May 1943, these were transferred to P Building (Fig. 44), and later two more TWX machines were added, with Army coders. In November 1945, the University of California canceled the contract on one of these machines, and at the end of 1946, the Army operated with two machines and coders.

Natural-Gas System. Early in 1946, the possibility of installing a natural-gas system, thereby eliminating monthly use of 300,000 gal of fuel oil and 1,500 tons of coal, as well as saving a little electric power, was investigated. The feasibility of such a plan was clear, and a contract was entered into with the Southern Union Gas Company of

Fig. 43. Technical Building T, looking southwest. This was the first building completed and in use in the Technical Area. The Administration and Theoretical Building, it contained offices, a photographic laboratory, a technical library, a document room, patent headquarters, and a concrete vault for classified material. The bridge across the road connected with a passageway between Gamma Building and E Building, not shown.

Fig. 44. View of office building, P Building. P Prime Building is visible through the guard tower at the left.

New Mexico and Texas, Dallas, Texas, to lay the necessary pipe to provide the Project three and a half million cubic feet of gas per day. This line consisted of 28 miles of 10-in. pipe additions to the main line and parallel existing lines near Farmington, New Mexico, approximately 130 miles distant, and the laying of 20 miles of 8-in. line from a point near Santa Fe to the Project -- about nine miles of which was laid by McKee with the rest laid by the Gas Company. It was also necessary to install one 400-hp compressor station at Bernalillo, New Mexico.

A contract to convert the existing heating and power facilities to natural gas was let to the International Manufacturing Company, Kansas City, Missouri. This work involved converting the two 1715-kW Nordberg generating units to use natural gas during off-peak seasonal demand periods only. This conversion was to be completed by about September 1947.

In December 1946, a 4.5-mile additional parallel 10-in. pipe line was laid close to Farmington, New Mexico, to increase the output to four million ft^3 per day. The total cost for this output, was estimated to be $704,000. A further plan for the Gas Company to lay other parallel lines and an addition to the compressor station was to eventually bring the output up to five million ft^3 per day. It was estimated that this would increase the total cost to $981,000.

The system was planned to ultimately include gas mains and laterals throughout the entire Project to furnish necessary gas for heating and cooking in most of the houses. Substandard houses and trailers were not considered in this group.

Progress

Unique construction problems impeded the contractors' progress. It was very difficult to obtain qualified skilled workmen, because of the isolation of the Project and because the immediate area was not a good labor market. Transportation of employees and building materials was over the difficult access roads. Housing for workmen was not adequate. Construction schedules could not be adhered to, for it was often necessary to start other jobs of equal or higher priority using the same crews, thus slowing down the original work or stopping it completely. The type of technical construction, the secrecy involved, and the terrific time pressure were other delaying factors. The Sundt Construction Company's original contract was 54% complete on February 2, 1943, approximately two months after construction started, but, because of the expansions, it was not 96% complete until April 25th. After that, increased construction requirements extended the original contract to December 15, 1943. Inasmuch as the McKee Company was not assigned a large amount of work at any one time, but was awarded additional construction as it became needed, it can only be stated that the company met the dates set with few exceptions.

The construction forces at the Project varied from time to time, because sometimes the work apparently was almost complete and the contractors cut down on their personnel; then at other times, when a number of urgently needed facilities were required, the contractor had to hire as many men as possible. Sundt's force varied from 250 to a peak of 3,000, whereas McKee's varied from 100 to 1,500, but was usually 700 to 1,000. Force Account crews were built up from 75 to approximately 1,800 men during the summer of 1945. Construction progress could be summarized by saying that completion dates and schedules were maintained for individual jobs only, not for the construction as a whole, largely because it was not practicable to forecast the entire program.

Costs

The following tabulation shows (in round numbers) the total cost of the Los Alamos Project, as furnished by the Cost and Accounting Section of the Fiscal Division.

Major Construction Contracts through Albuquerque District	$ 9,315,000.00
Design and Engineer (Kruger)	175,000.00
Major Contracts through Manhattan Engineer District	17,519,000.00
Design and Engineer (Kruger)	521,000.00
Cost of Utilities (Including connection of new power line to New Mexico Power Company)	6,849,000.00
Design and Engineer (Kruger	48,000.00
Black and Veatch)	164,000.00
Force Account (Estimate)	6,000,000.00
Operation and Maintenance	1,416,000.00
Real Estate	5,000.00
Warehouse Inventories, Salaries, Miscellaneous	15,458,000.00
Grand Total	$57,880,000.00

The Force Account costs include some operation and maintenance, because hired labor crews worked in maintenance as well as construction, and no line of demarcation was established in maintaining costs until the end of 1945 when the Cost and Accounting Section was formed.

Because the construction work originally planned for the Project was relatively small, the original contractor did not set up a very large plant or organization to handle the work. The increased work came in small amounts, and the construction contractor did not employ the required personnel or build plant facilities that would normally have been provided for a job of the size that finally developed.

Another reason for the high cost of individual items, relative to such items at other localities, was the Project's isolation. The highway to the site was not improved until after Sundt had completed the first construction work. This road was very narrow, had steep grades, poor alignment, and several switchbacks, and was very dusty. All materials and equipment for the Project had to be hauled from the railhead at Santa Fe, approximately 46 miles over this bad road. McKee's work was also costly compared with work at better-located jobs. However, McKee's work generally cost less than that of the original contractor. This is attributed to the fact that the access road was in better condition.

Furthermore, many complex mechanical installations were required. Generally these installations comprised most of the work in the technical buildings and outlying sites. Completion of a building was often delayed by slow or uncertain delivery of critical technical items. All construction work was conducted on the basis that the crews would be removed from the job whenever required by the University of California, and construction crews often lost considerable time because of this condition. The nature of the work militated against furnishing design information that would remain unchanged during construction. After it was determined that additional facilities were required, the time allowed for construction of the buildings or facilities was very short. Therefore construction often began when only a small part of the plans was completed. The many changes made during construction added to the cost.

It was originally assumed that the Project could be kept to a maximum of 300 persons, and it was upon that figure that plans for utilities and housing were based. This assumption proved woefully inaccurate. Had the ultimate size of the Project been fully anticipated, the various construction contractors could have planned their housing, messing facilities, and utility programs more accurately, thus obviating much of the labor difficulties, which, in turn, would have lowered construction costs.

V. COMMUNITY ADMINISTRATION AND OPERATION

General

Headquarters at this station was directly responsible for Post security, handled first by the MP Detachment, later through the Intelligence Officer. This included guarding all entrances and outlying sites and patrolling the roads. Post operation and maintenance were also the Army's responsibility. Furnishing lights, water, and sewage facilities; furnace firing; garbage hauling; building and road maintenance; keeping the Post clean; and similar functions were handled primarily by the men of the Provisional Engineer Detachment of the 8th Service Command. The Post operated all mess halls, using civilian and military personnel supervised by men of the Engineer Detachment. Commissary and Post Exchange, operated by the Army, were the only means for Post residents to secure food supplies and other necessities. Another direct Army responsibility was to provide furniture and similar articles for all dormitories and, to a large extent, to families living in apartments. The Post operated the Station Hospital and staffed it with Medical Officers; however, all nurses and later two female doctors (married to men employed on the Project, and working part time) were on the University of California payroll. This was mainly because nurses could not be hired for the salary set up by Civil Service, nor were Army nurses available.

These responsibilities were handled by the Army until early in 1946 when it became apparent that the military units would continue to shrink and become unable to perform these services satisfactorily. After much discussion, a new company was formed on April 1, 1946, as a municipal organization to take over these services. The Zia Company, as it was called, entered into a cost-plus-fixed-fee contract with the Government on that date. They took over the operation of hospital, schools, transportation, housing, utilities, maintenance, and other service functions, with the civilian personnel formerly engaged in these duties for the Government transferred to Zia and carried on their payroll. The Zia Company officers were Robert E. McKee, President, and Jack Brennand, Vice President. Resident Manager was Elmo Morgan, formerly Lt. Col. Morgan of the Albuquerque District Engineers Office.

Housing

Civilian Housing. Housing for civilians was provided primarily for the University of California scientific personnel. As many as possible of the personnel working for the Commanding Officer in a service capacity were to commute from Espanola, Santa Fe, or other nearby communities. As the Project grew, it became more and more difficult to find enough qualified service personnel who were willing to commute. Therefore, key personnel were allowed to be housed on the Site, with the provision that, should scientific personnel need the housing, the service personnel would have to find living accommodations elsewhere. Navy officers were considered in the same category as scientific personnel and accorded the same housing privileges. A minimum number of Army Officers were allowed family housing on the Project, and it was understood that should housing become essential for scientific personnel, these officers would be asked to vacate on 30 days notice.

The number of people housed at the Site was always kept at the minimum commensurate with the workload. Had adequate housing been available for service personnel, it would have been much easier to provide the necessary services. The lack of housing caused great personnel turnover, and it was very difficult to induce skilled mechanics or qualified engineering personnel to work at the Site. Housing was kept to a minimum because each new family multiplied the service personnel needed for maintenance. The limited water, electric power, and sewage-disposal facilities, and the transportation of fuel to the Site were additional problems. Hospital, messing, commissary, and other public facilities required expansion in direct proportion to the number of families housed.

Family housing (Figs. 45-64) at the end of 1946 included apartments, duplexes, and houses for 617 families, as follows.

332 Sundt Apartments
56 Morgan Duplexes
100 McKee Prefabricated Houses
107 Hanford Prefabricated Houses
11 Metal Prefabricated Houses
9 Master Houses
2 Western Development Houses

A further housing program was to provide 300 additional Western-Type units, permanent structures of stucco and concrete block. Also 240 houses from Ft. Leonard Wood, Missouri, were shipped in to be reassembled early in 1947. Another addition of 39 metal prefabricated houses was planned for use by the first three grades of Noncommissioned Officers.

The 617 available units did not include 16 Ranch Houses which were remodeled Ranch School structures, 51 winterized hutments, 47 Government-owned standard trailers, 25 Government-owned expansible trailers, 56 Pacific Hut apartments, 30 National Hut apartments, and approximately 250 privately owned trailers (Figs. 65-70). The Ranch Houses, hutments, and trailers were inferior to the 617 apartments, duplexes, and houses constructed under Government contract, some of which, in turn, could not be considered permanent housing.

Dormitories (Figs. 71-75) were provided for single male and female civilians, and there were a few dormitories for married couples without children. All of the better dormitories were to house scientific personnel. Some rooms were made available to key personnel of the service organization working under the Commanding Officer. There were dormitories of cheaper

Fig. 45. Efficiency-apartment building.

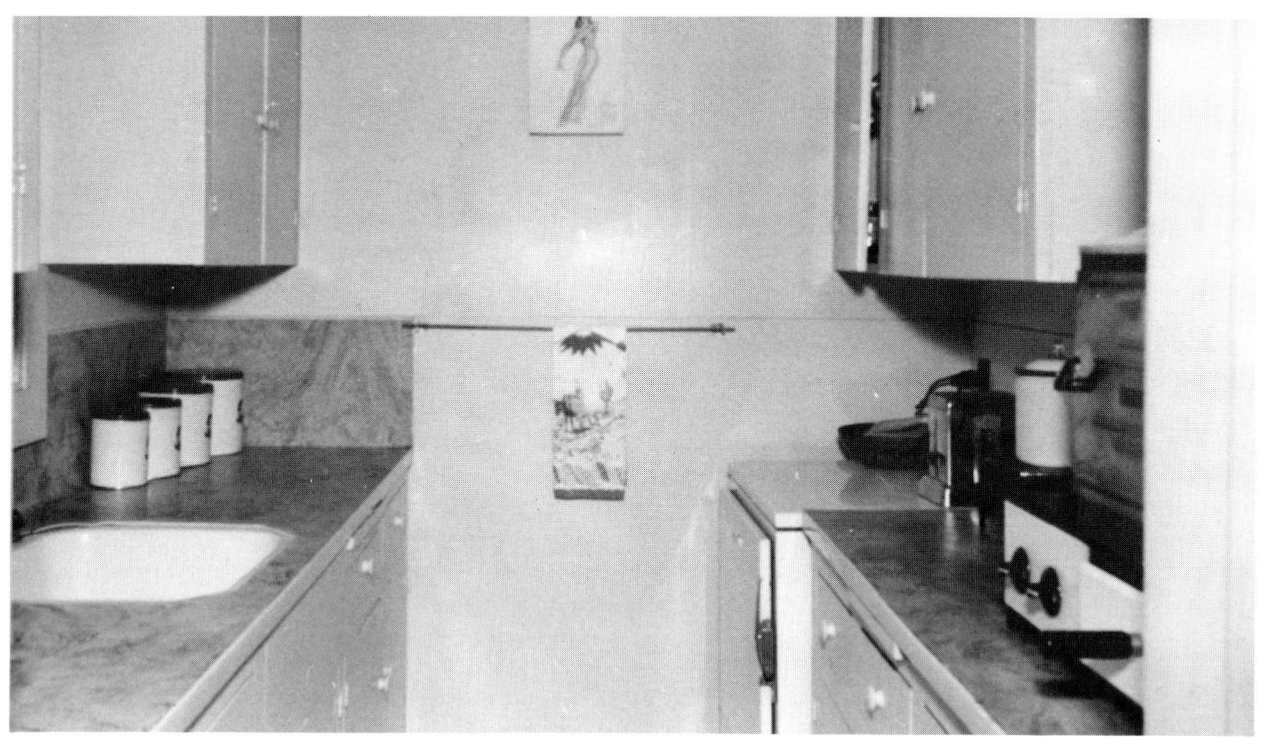

Fig. 46. One type of kitchen in an efficiency apartment.

Fig. 47. Living room in an efficiency apartment.

Fig. 48. Sundt-built, one-bedroom duplex.

Fig. 49. Living room of a Sundt duplex.

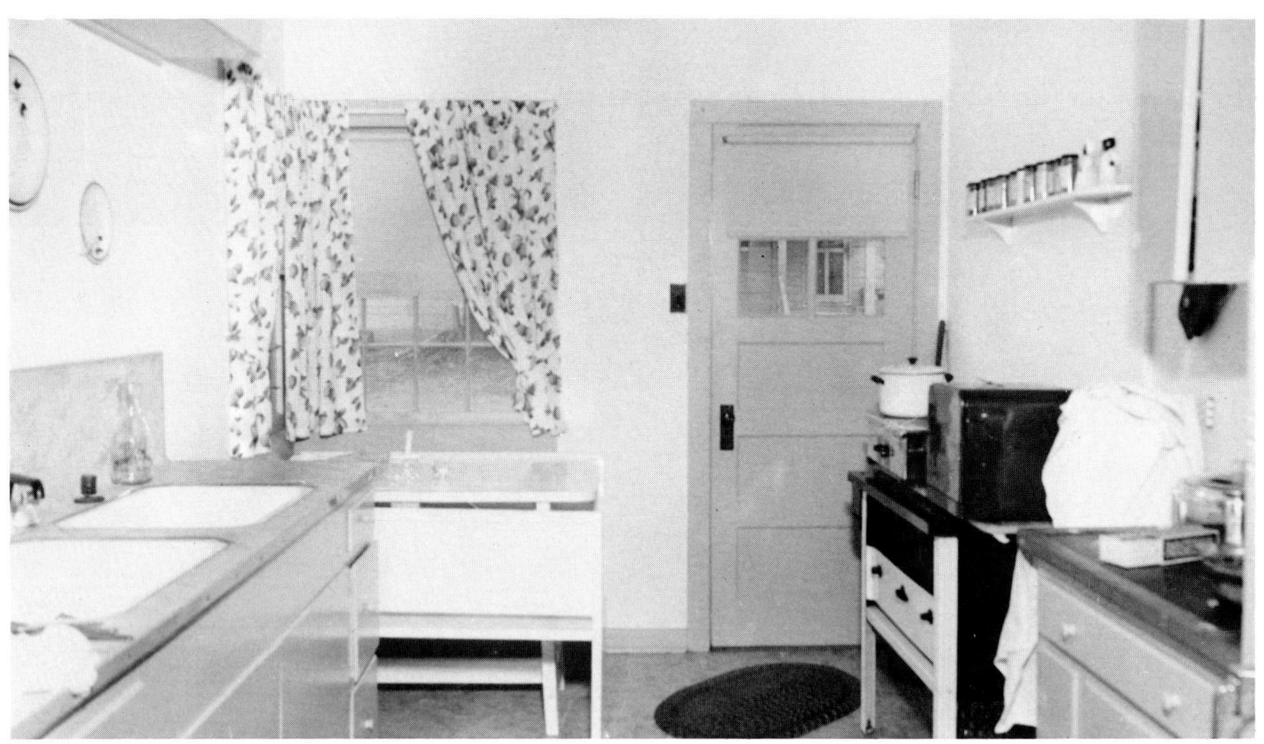

Fig. 50. Kitchen in a Sundt duplex.

Fig. 51. Two-bedroom-type apartment building.

Fig. 52. Living room in a two-bedroom Sundt apartment.

Fig. 53. Kitchen in a two-bedroom Sundt apartment.

Fig. 54. Apartment building with three-room apartments.

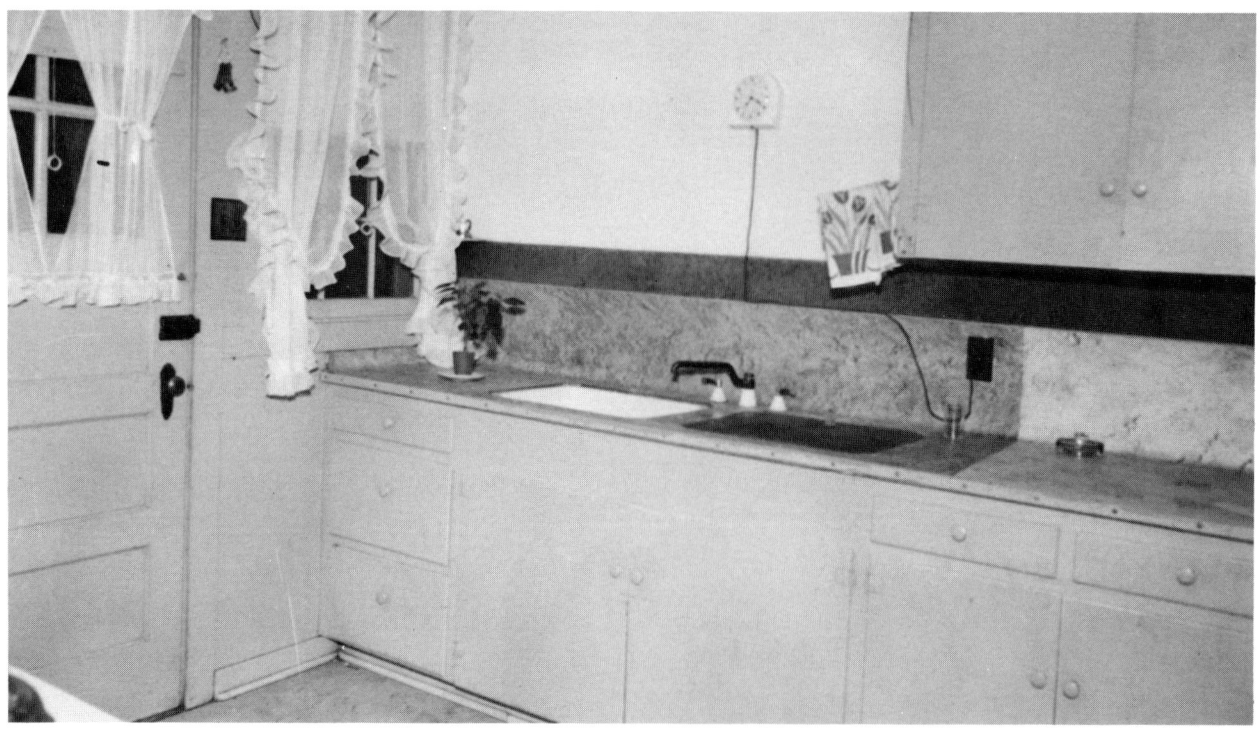

Fig. 55. Kitchen in a three-bedroom Sundt apartment.

Fig. 56. Sundt four-room apartment building.

Fig. 57. Morgan duplex houses.

Fig. 58. Living room in a Morgan duplex.

Fig. 59. Kitchen and dinette in a Morgan duplex.

Fig. 60. Kitchen in a Morgan duplex.

Fig. 61. McKee flat-roofed housing.

Fig. 62. Pasco, or Hanford-type house with two bedrooms.

Fig. 63. Three-bedroom Pasco house.

Fig. 64. A permanent Western house.

Fig. 65. The Los Alamos Ranch School Arts and Crafts Building, used as a Master House for the Laboratory Director. The Jemez Mountain Range lies in the background.

Fig. 66. This Ranch School Master House, known as Spruce Cottage, was originally used as the WAC headquarters. Later it was used for three apartments, then divided into two master apartments.

Fig. 67. Hutments.

Fig. 68. Pacific hutments.

Fig. 69. Expansible Government trailers.

Fig. 70. The Trailer Area, showing privately owned trailers in the foreground and Government trailers in the middle distance.

Fig. 71. Men's dormitory.

Fig. 72. Men's dormitory--capacity 20 men. Semiprivate baths were the feature of this type of dormitory. Only four of these, two for women and two for men, were built.

Fig. 73. Example of a men's dormitory room.

Fig. 74. Dormitory dayroom.

Fig. 75. Women's dormitory.

construction for mess attendants, furnace firemen, janitors, hospital attendants, and other low-salary employees. At the end of 1946 there were 36 dormitories with approximately 1253 living quarters, and 55 barracks providing another 1496 individual units.

Better housing was provided, mainly for visitors, in Fuller Lodge, Fuller Lodge Guest Cottage, and the Big House (Figs. 76-78), stone and log structures from the Ranch School. These quarters were available for transients and some of the senior scientific personnel who were single.

Approximately 118 hutments, 16 by 16 ft, were erected by the original construction contractor to house his tradesmen and those employed directly by the service organization in Force Account work. These hutments were used from time to time, depending upon the amount of work being done.

Robert E. McKee, Construction Contractor, furnished the labor to build an H-shaped dormitory to house about 93 of his key personnel. It was built from CCC sections furnished free by the contractor on the basis that when his company finished work the structure would be available for the service organization.

Housing control and administration of both apartment and dormitory facilities were under the University of California until February 1946, when the Army assumed control. The primary school of the Ranch School was established as the Housing Office, where all the administration work was done and all records were kept. This was later converted into the Technical PX, and another of the original Ranch School buildings was used for the Housing Office. Before the Army assumed control, a liaison officer worked with the University of California housing authority on all matters requiring cooperation or approval of the Commanding Officer. Such matters included rental rates, building maintenance, quotas of housing facilities to be made available for the Commanding Officer's employees, and requests for and provision of additional housing. The Housing Office also provided maid service, express deliveries, etc., until the Zia Company took over in April 1946.

Rental rates for family quarters were established in accordance with Orders B of the War Department (see letter, page 59) and were approved by the District Engineer, Manhattan Engineer District, in February 1943. Rates were based on Civilian Personnel Regulations

Fig 76. Fuller Lodge, the headquarters of the Los Alamos Ranch School, served as the hotel and main dining room for the Project. Part of the Guest House and Residence is shown at the right.

Fig. 77. Guest Cottage and Residence. The stone section is the original Ranch School Guest Cottage, comprising the Blue and Brown Rooms, in which were housed the more prominent visitors to the Project. The log structure is a private residence, also an original school building. In the right background is Spruce Cottage.

Fig. 78. Big House, originally a Ranch School dormitory, was used for the library, Chaplain's Office, and Red Cross headquarters. A machinists' barracks is in the background.

WAR DEPARTMENT
United States Engineer Office
Manhattan District
P. O. Box 42
Station F
New York, N. Y.

P. O. Box 1539
Santa Fe, N.M.
Feb. 10, 1943

Subject: Establishment of rates for quarters for the operating contractors' employees on the Zia Project.

To: The District Engineer, U. S. Engineer Office Manhattan District, New York, N. Y.

1. It is requested that authority be obtained for the establishment of rates for the use of Government quarters by operating contractors' employees at the site. In view of the fact that these contractors' employees are going to be working and living in close proximity with the Government employees, and because they are occupying the same type of quarters, and because they must of necessity live on the post, it is recommended that the rates authorized for Government employees in Orders B of the War Department, dated January 15, 1943, be also applied to the contractors' employees.

2. The main reasons for recommending the same rates for contractors' employees as for Government employees are twofold; it is a simple method of handling the situation and it will forestall any possible trouble in case different rates were established for each group.

3. It is further requested that authority be obtained to charge the rate of $1.00 per lodging per night for authorized guests, other than Government and contractors' project employees.

4. In view of my attempt to utilize the services of the wives or children of Government employees on the post so as to keep the population at a minimum, it is also requested that authority be obtained to waive the charging of quarters to an employee when one employee living in the same quarters is already having the value of quarters deducted from his salary. It does not appear proper to charge the wife of a contractor's employee for quarters while living in the domicile of her husband and it would be unfair and would cause dissension to charge a Government employee's wife for quarters occupied with her husband.

/s/ J. M. Harman
J. M. HARMAN
Lt. Col., Corps of Engineers

promulgated by the General Accounting Office. The rates, based on yearly salary, were as follows.

Yearly Salary		Monthly Rent
$ 0	$ 840	$ 5.00
841	1080	8.00
1081	1560	10.00
1561	1920	12.00
1921	2200	15.00
2201	2600	17.00
2601	3100	23.00
3101	3400	29.00
3401	3800	34.00
3801	4400	42.00
4401	5200	50.00
5201	6000	59.00
6001	and up	67.00

Utility fees were comparable to those at other Projects at which Government housing was furnished. During the Spring of 1944, the utility cost was estimated and found to exceed the actual charges by some three to five times. After considerable reviewing, the utility charges were not changed for the following reasons.

Actual cost greatly exceeded rates being charged in normal cities and communities. This was because of the Project's isolation, the inefficiency of the local labor, and the high cost of transportation from the nearest railhead in Santa Fe.

An adjustment in utility charges would have caused considerable unrest in the civilian population.

Good living quarters at a reasonable rental and utility rate were an inducement to obtain civilian personnel to staff the Project adequately.

Actual utility charges were as shown in the following table.

Accommodations in the efficiency apartments, one bedroom and a living room, with a tiny bath and kitchenette, were considerably poorer than those in the other apartments with one or more bedrooms. The District Office authorized a flat 15% rental reduction for efficiency apartments in December 1943. The accommodations in the remodeled ranch houses and winterized hutments

UTILITY CHARGES

	Efficiency Apartments	Early Morgan and McKee Apartments								
		3-Room			4-Room			5-Room		
Water	$0.50	$0.50	$0.50	$0.50	$ 0.60	$ 0.60	$0.60	$ 0.70	$ 0.70	$ 0.70
Refrigeration	1.15	1.15	1.15	1.15	1.35	2.35	1.35	1.35	1.35	1.35
Electricity	3.00	2.00	3.00	3.00	2.50	2.50	2.50	3.30	3.30	3.30
Heat	3.00	4.00	3.00	3.00	5.00	5.00	5.00	6.00	6.00	6.00
Janitor	1.00	2.00	2.00	None	1.50	1.50	None	1.50	1.50	None
Total	$8.65	$9.65	$9.65	$7.65	$10.95	$10.95	$9.45	$12.85	$12.85	$11.35

	Ranch Houses and Winterized Hutments				Master Houses	Guest Houses
	2-Room	3-Room	4-Room	5-Room		
Water	$0.50	$0.50	$0.60	$0.60	$ 0.70	$1.00
Refrigeration	1.15	1.15	1.35	1.35	1.35	per day,
Electricity	1.50	2.00	2.50	2.75	3.40	including
Heat	2.50	3.00	3.50	3.75	6.00	utilities
Janitor	None	None	None	None	3.00	
Total	$5.65	$6.65	$7.95	$8.45	$14.45	

were considerably below scale (see letter, next page), and they were also reduced 15%.

Rent for both Civil Service and University of California employees was based on monthly salary, as shown above. That for Civil Service employees was figured on their basic 40-h per week salary, although they were working 48 h a week. That for University of California employees was reckoned on their basic salary which covered a 48-h week, until March 1945, when it was also figured on a 40-h work week and reduced. When the work week was reduced from 48 to 40 hours without change in pay, there was a corresponding increase in rent for the University of California employees.

Trailer space with utilities was provided for private trailers at $5 per month. The 47 Federal Housing Administration trailers (Government-owned standard trailers) rented for $28.00 a month, including all utilities, and the 25 Government-owned expansible trailers rented for $33 a month, including all utilities. These rates were determined by the charges made at Dallas, Texas, by FPHA for the same accommodations. The 16- by 20-ft Pacific Hut Apartments were considered comparable to the standard trailer units, and also cost $28.00 a month.

Quarters for single personnel living in dormitories or barracks rented at the following rates.

No. in One Room	Monthly Rent
1	$ 13.00
2	8.00
3 or more	6.00

Maid and linen service cost $2.00 per person per month.

Military Housing. Troop housing (Figs. 79-83) was to have been 40 ft^2 per enlisted man and 50 ft^2 per enlisted woman, as provided by War Department circulars on the subject. However, the rapid growth of the military units and the lack of barracks space made it impossible to adhere to that policy. Structures used were Theatre of Operations type, or modified mobilization type in the case of WAC barracks buildings. Units housed were the Special Engineer Detachment, WAC Detachment, Military Police Detachment, and the Provisional Engineer Detachment. Some barracks not in use by the Army were used for civilian barracks or converted to dormitories.

Building Maintenance and Repair

Community structures were generally maintained and repaired by Force Account crews. The normal procedure was to phone the maintenance office with a request for service which was then sifted for approval and issued to the proper trade foreman. Routine calls were handled according to priority need. Emergency calls were handled by crews maintained for 24-h service. Crews were larger than those usually required for a similar installation, because of the haste of original construction and the isolation of the Project.

Further work was necessary because of the considerable shifting of residents in the housing area. This was done by Government-employed personnel using Government-owned equipment, because it was not desirable under existing security regulations to have commercial moving concerns enter the Project. A crew went into vacated apartments to clean, repaint, and make minor repairs, to provide suitable quarters for the next occupant. This was considered a part of the Government maintenance and was done without charge to the occupants.

Ice Delivery. Another service, ice delivery, was established because of the difficulty in supplying electric refrigeration. Ice from the ponds in the area was cut in the wintertime and stored in an ice house, for later sale to residents of facilities that did not have electric refrigeration. This project was self supporting. After addition of the trailer area, ice had to be hauled to the Project from Santa Fe. Ice was delivered to trailers and other dwellings, and the occupants were billed for this service.

Heating. Heat in the apartments, dormitories, and duplex units was maintained by a force of 60 to 70 civilian janitors, including 20 to 25 wood and coal delivery men, approximately seven months out of the year. The furnace-firing areas were divided into 16 districts with a man for each district for each of two 8-h shifts per day. Enlisted personnel who had approximately four districts each supervised these shifts. There was considerable difficulty because the firemen were inexperienced, many could not speak English, and most of them could not work on a permanent year-round basis. Therefore, the turnover was great, and the problem of firing the furnaces caused considerable dissatisfaction among the occupants of the apartment buildings.

131. EIDM CO

7 January 1944

SUBJECT: Rent Reduction for Improvised Quarters

TO: The District Engineer, Manhattan District, P.O. Box 1111, Knoxville, 7, Tennessee

1. Some time ago your office authorized a rent reduction for efficiency 1-bedroom apartments amounting to 15% of the rental.

2. We are now engaged in construction of a camp consisting of winterized hutments converted into apartments of various sizes for the maintenance personnel on the Post and are also converting old buildings, such as the former blacksmith shop, tire repair shop, warehouse, etc., into apartment buildings for maintenance and construction personnel employed by the Post Engineer. Since these quarters are not on the same level with the new apartments constructed at this post, it is requested that permission be granted to reduce the rents in instances where in the opinion of the undersigned quarters assigned an individual are below the standard of quarters assigned to other personnel with the same size families. For the purposes of uniformity, it is recommended that this reduction be 15%, in line with the previously authorized reduction described above.

/s/ Whitney Ashbridge
WHITNEY ASHBRIDGE
Lt. Col., Corps of Engineers
Commanding

Fig. 79. Metal prefabricated house for enlisted military personnel.

Fig. 80. WAC Barracks, 76-woman capacity.

Fig. 81. Enlisted men's barracks, type B-A-T.

Garbage Disposal. Garbage and ashes were collected from each dwelling daily by Government-employed personnel using Government-owned equipment. Wet and dry garbage from family quarters was burned in an incinerator (Fig. 84) built for the purpose. Ashes were collected and used for fill or for surfacing low-cost roads. Edible garbage was removed from the mess halls and sold to a near-by farmer for his hogs.

Self-Help Laundries. There were six laundries, equipped with domestic-type electric washing machines, mangles, and hand irons (Fig. 85). A superintendent, charwoman, and janitor were required to operate each installation. Any resident was eligible to use the equipment and was charged by the hour. These central laundries were established because private facilities were not available.

Eating Places

Eating places and the commissary were of prime importance on such an isolated Post where all meals and staple food had to be provided. It was the policy here, as at Projects where only Government eating facilities were available and where Government employees were involved, to provide meals at practically food cost, with the Government subsidizing the eating establishments on that basis. Therefore, the cost of operation showed an expected loss. However, after September 1946, prices at the various restaurants were raised considerably to partially offset this. (See letters, pages 68 through 70.)

Fuller Lodge. In February 1943, Fuller Lodge, one of the Ranch School buildings, was taken over by the Government and operated under Civil Service for transient housing and messing of Post and Technical Personnel. The Lodge had one large dining room, two stories high, with a balcony and two mammoth stone fireplaces practically covering the entire north and south ends of the room. This room had an approximate seating capacity of seventy. A smaller private dining room, called the "Curtis Room," seated twenty. This and a similar room upstairs had stone fireplaces. Nine bedrooms in Western style were available. From March through July 1943, about 4000 meals per month were served there. This gradually increased to about 13,000 meals per month. The original

Fig. 82. Interior of enlisted men's barracks.

group of WAC's was served at Fuller Lodge, but when the group became too large, the new arrivals ate at North Mess. Later when the WAC's moved to the Western Area, they ate at the West Mess. The Lodge had a group of regular guests who paid $60.00 per month for meals, but as this group was much smaller than the actual capacity, other site residents could have casual meals at the Lodge, normally making reservations a day in advance. Originally the board at Fuller Lodge was $25.00 per month. Charges through 1944 were $0.50 for breakfast, $0.65 for lunch, and $1.00 for supper, raised on October 1, 1946 to $0.75 for breakfast, $0.85 for lunch, and $1.15 for supper. Fuller Lodge Guest Cottage, part of the Lodge operation under Mr. H. M. Acher, had two bedrooms, one fireplace, and private baths, used for official visitors only.

North Mess. In April 1943, North Mess (Fig. 86) opened to accommodate the civilians on the Project, and 675 meals were served the first day. On the third day, expansion was started to accommodate the increasing patronage. During the second week of operation, 5800 meals were served, the third week, 6300 meals. Other structure additions were made in July and October 1943, and in January 1945, raising the seating capacity to 400 persons. The patronage during this period increased to 58,250 meals per month, and the staff grew to 1 Mess Sergeant, 12 cooks, 20 kitchen helpers, and 32 mess attendants.

Fig. 83. Bachelor Officers' Quarters, a semibarracks using space heaters. Theater No. 1 is to the left.

Charges for meals were $22.50 or $25.00 per month, depending on attendance over week-ends, or $0.40, $0.50, and $0.65, for breakfast, lunch and supper, respectively. The rate in September 1946, was $0.65 for each meal, or $10.50 for 21 meals.

West Mess. West Mess (Figs. 87-88) so named because of its location at the west end of the community, opened in September 1943. It served the civilians living there and also the WAC Detachment until the Spring of 1945, when their own mess hall (West Cafeteria) was completed. The original staff of 1 cook, 1 helper, and soldier KP's grew to 1 Mess Sergeant, 2 bakers, 1 butcher, 12 cooks, 12 cook's helpers, and 38 attendants and charwomen. In January 1944, all cooks were military because civilian cooks could not be obtained nor could living quarters be provided. Late in 1946 West Mess was converted into a completely military mess for the combined military detachments. While it was open to civilians, the charges per meals were the same as those at North Mess.

A small mess with a seating capacity of 250 was established at S Site in March 1945, under the management of the West Mess. This was used for the convenience of S Site personnel only.

WAC Mess, or West Cafeteria. This building was completed in the Spring of 1945. Its original staff was six cooks and one Mess Sergeant. During the first month's operation it provided mess for the WAC Detachment only; then civilian and military personnel alike were served, until October 1946 when the WAC's were inactivated. Then it became West Cafeteria, with the same price level as North Mess. Its serving capacity was 158 persons.

East Cafeteria. In March 1945, East Cafeteria (Fig. 89) opened under the management of Mr. H. M. Acher. It was more decorative, had nicer furnishings, and provided a wider range and variety of excellent foods than did the Mess Halls, to give Project personnel better meals in a more restful environment. In this respect it was definitely a morale factor. It had a seating capacity of 400 and served resident and transient, military and civilian personnel. All meals were on an individual cash basis. The Cafeteria was readily accessible to most of the dormitory residents, and their

Fig. 84. Incinerator, 5-ton capacity.

patronage was large. East Cafeteria was operated on an a-la-carte basis, with prices based on food costs and certain overhead costs.

Military Messes. In addition to these eating establishments, a mess hall for the Military Police Detachment was constructed early in 1943. It also provided meals for the Provisional Engineer Detachment until that organization was inactivated in July 1946. This Military Mess was closed when the military organization was set up in the Western area.

The Post Exchanges, also provided eating establishments; in fact, the Tech PX at one time operated 24 hours a day.

Commissary

The Commissary was opened in March 1943. The opening stock was taken over from the Ranch School Trading Post, and an entire stock was purchased from a merchant in Santa Fe who was closing out his store. At first the intention was to sell only to employees actually living on the Project; however, it was the same old story of Project expansion beyond the original plan. Greater numbers of people who lived off the site were hired, and, because they could not reach other food stores during the working day, the commissary opened its doors to them as well. As the Project continued to grow, the original installation had to be expanded, and storage facilities for perishable and nonperishable foods, as well as additional store space, were built. The warehouse space was urgently needed because of the isolation and the long haul involved.

The necessity of maintaining good morale among the civilian population in this isolated community made it desirable to stock a large number of items ordinarily not provided by Army Regulations, through the Quartermaster Depot, whose main function was to supply military messes. Permission was granted in August 1944 to carry "Unauthorized Items." (See letter, page 74.) These items were procured in accordance with law, and were sold at prices that precluded any loss to the United States.

Various personnel problems entered into the Commissary operation. The limited Civil Service salaries and the limited housing made it necessary to supplement the staff of butchers, warehouse men, etc., with military personnel. The following tabulation shows the personnel situation.

Date	Military Staff	Civilian Staff
April 1943	0	8
April 1944	11	15
March 1945	14	32
March 1946	12	79
December 31, 1946	10	83

All accounting was originally handled by the Fiscal Section of Project Headquarters. In the fall of 1944, an accounting section was established in the Commissary to receipt, issue, and requisition according to Quartermaster procedure. There were minor deviations because of local

ARMY SERVICE FORCES
UNITED STATES ENGINEER OFFICE
P. O. BOX 1539
SANTA FE, NEW MEXICO

In reply
refer to: EIDM FF-4

22 AUGUST 1946

SUBJECT: Mess Operations - Site Y

TO: The Commanding General
Manhattan Project
P. O. Box 2610
Washington 25, D. C.

1. Audit Reports for the period 1 January 1946 to 30 June 1946 disclose an operating loss at Fuller Lodge, East Cafeteria, and North Mess of approximately $77,970.51. Actual figures for two quarters at the North Mess and one quarter at East Cafeteria and Fuller Lodge are $62,970.51. Based on past experience, Fuller Lodge and East Cafeteria will lose approximately $15,000.00 for the second quarter.

2. An analysis of the operating losses of these messes indicate that the following facts have contributed to the net operating loss:

a. Sale of Meal Tickets at North Mess for $22.50 per month and $50.00 per month at Fuller Lodge has resulted in substantial losses per meal each month. In this connection it is not mandatory that meal tickets be furnished civilian employees for subsistence at Government operated messes when adequate Post Restaurant facilities are available as outlined in Civilian Personnel Regulation No. 125.

b. Also the high overhead cost of operation of these messes is occasioned by the fact that only a 40 hour tour of duty is established for employees, resulting in the necessity of having two or more shifts to efficiently operate these messes.

3. Operation of eating establishments at this project is a necessity, and it was mandatory during the period of greatest manpower shortage to subsidize mess halls and resort to other means of making employment at this remote project attractive. However, this project's peak manpower shortage has been passed, and it is believed that employment problems no longer constitute justification for mess hall subsidization, particularly North Mess.

4. Approval has been granted the Engineer Exchange to increase price of meals in order to meet increased food costs, however, no increase has been made in the price of meals served in the above messes.

5. If action is approved to correct the discrepancies listed above the proposed procedure will be as follows:

a. Increase the price of meals served.

b. Discontinue sale of $22.50 meal tickets at North Mess and $50.00 meal tickets at Fuller Lodge.

c. Place messes on a concession basis.

6. It is requested that opinion be furnished whether or not subsidization of mess halls should be continued and/or whether corrective action as per above, should be withheld until the Civilian Commission has taken the matter under advisement.

L. E. SEEMAN
Colonel, C. E.
Commanding

5 Incls:
Cost Stmt Mess Oper.
North Mess, Jan. Feb. Mar.
Apr. May and June
Cost Stmt Mess Oper.
Fuller Lodge, Jan. Feb. Mar.
Cost Stmt Mess Oper.
East Caf. Jan. Feb. Mar.
Recap of Net Oper Losses from
Mess Oper (E. Caf. - N. Mess -
Fuller Lodge) for qtr ending 31 Mar.
Summary of Net Oper Losses N. Mess
for qtr ending 30 June.

Subject: Mess Operations--Site Y

1st Ind.

Manhattan Project, P. O. Box 2610, Washington, D. C., 27 August 1946.

To: Commanding Officer, U. S. Engineer Office, P. O. Box 1539, Santa Fe, New Mexico.

You are authorized to take any steps which, in your opinion, will improve the financial condition from the standpoint of the Government and will not be detrimental to the overall efficiency of the project.

L. R. Groves

L. R. GROVES
Major General, USA

Incls. n/c

Fig. 85. A self-help laundry.

Fig. 86. North Mess Hall.

Fig. 87. West Mess, or SED Mess Hall.

Fig. 88. Interior of West Mess.

Fig. 89. East Cafeteria

conditions. One was a 10% charge on cash purchases, to offset additional services and losses, such as delivery service, milk bottle losses, and check-cashing costs. Another was handling larger sums of money than Army Regulations stipulated, because security would not permit employees to have bank accounts in Santa Fe or nearby communities. As no other organization could handle enough funds to cash pay checks for the Project, the Commissary was the logical organization to perform this service. In October 1944, authority was established to carry approximately $30,000.00 for check cashing. Arrangements were made at the same time to install an adequate safe for this money. (See letters, pages 75 and 76.) In November 1946, the amount was changed to $85,000.00.

The following outlines the growth of Commissary sales over a two-year period.

Date	Cash Sales	Sales to Organizations
March 1943	$ 672.42	$ 479.31
March 1944	20,829.20	16,612.46
March 1945	39,892.00	56,534.00
March 1946	83,000.00	10,310.00
December 31, 1946	92,824.94	66,952.70

In June 1946, the policy of selling only to residents was again established, for it was decided that the real emergency was over and food supplies were becoming more available in local markets. Commissary cards were then issued only to those living on the Project.

Post Exchanges

The Post Exchange was activated in June 1943. The only facility of this sort until then was an MP branch exchange for military personnel, operated under the Main Exchange of Bruns General Hospital. The sum of $5,000.00 was borrowed from the Army Exchange Service, to purchase merchandise and defray expenses until the Exchange could handle its own obligations. This was paid back in monthly installments, the last payment being made in June 1944. The original exchange comprised the MP Exchange, the Barracks Bar, the Trading Post, and a General Store. The Barracks Bar was for Sundt Construction Company employees and closed in December 1943 after only a few months operation. Later it was reopened as the Noncommissioned Officers Club.

EIDM CO 12 August 1944

MEMORANDUM TO: Files

Because of the peculiar nature of the project at this Post and the necessity for maintaining good morale among the civilian population, it is considered necessary to carry in stock at the Commissary more items than are authorized by Army Regulations. The isolation of this Post and the security regulations are also contributing factors which prevent Post residents from access to civilian sources of supply.

This is made a matter of record in accordance with instructions from higher authority that we should put such a statement in the files in the event it became necessary to break Army Regulations for the proper prosecution of the project. It is considered by the undersigned necessary to carry a large variety in order to satisfy civilian personnel and so they may put forth their best efforts for the furtherance of the project.

/s/ WHITNEY ASHBRIDGE,
Lt. Col., Corps of Engineers

A TRUE COPY:

Edith C. Truslow

EDITH C. TRUSLOW
2nd Lieut. TC

EIDM ACO 13 October 1944

Subject: Excess Cash on Hand

To: The Commanding Officer, U.S. Engineer Office, P. O. Box 1539, Santa Fe, New Mexico.

1. In accordance with TM 10-215, Section V, Paragraph 43, the maximum allowable amount of cash to be retained at the Commissary at any one time is $200.00.

2. It is requested that authority be granted for the Commissary to retain the cash from the daily sales and collections for a period of 10 days, i.e., on the 10th, 20th and the last business day of each month, in order that a sufficient amount of cash will be on hand for check cashing purposes.

3. If authority is granted for retention of cash for the periods mentioned above, there will be in the safe approximately twenty thousand to thirty thousand dollars at one time. The safe in the Commissary is the safe taken over from the Los Alamos School and is no protection for that amount of money. Adequate facilities should be installed in the Commissary to safely handle this money.

VIRGIL F. DEAN
Captain, Corps of Engineers

A TRUE COPY:

Edith C. Truslow

EDITH C. TRUSLOW
2d LIEUT. TC

EIDM CO 1st Ind. WA:MJG

U.S. Engineer Office, P. O. Box 1539, Santa Fe, N.M., 16 October 1944

TO: Captain Virgil F. Dean, P. O. Box 1539, Santa Fe, N.M.

1. Approved in view of the necessity for providing check cashing facilities at this isolated project.

2. It had been suggested by the undersigned that you could use the large safe formerly used by the Intelligence Office, now in the Post Engineer Office, but it is understood that you and Lt. Maguire have examined the safe and consider that it offers no more security than the one now used other than its greater size. Efforts will be made, therefore, to obtain a more secure safe from the District, surplus lists, or such other source as may prove available.

WHITNEY ASHBRIDGE,
Lt. Col., Corps of Engineers
Commanding

A TRUE COPY:

Edith C. Truslow

EDITH C. TRUSLOW
2d Lieut. TC

The Trading Post, (Fig. 90) was a small log structure, with a 29- by 30-ft selling space and a small room in the rear. Soft drinks, beer, ice cream, tobacco, plate lunches, and hamburgers were served. It was the center of purchasing activity and much too crowded. In October 1943, the entire operation was exchanged with the operations in a larger room in Building T-7, which had been functioning as the General Store, selling drugs, clothing, and gift items; distributing laundry and dry cleaning; and handling mail. This reversal of operations made the Trading Post the General Store, and Building T-7 became the Service Club, carrying sundry Post Exchange items as well as serving meals. The Service Club (Fig. 91) also added a barber shop and beauty parlor operated by the Post Exchange.

Another Post Exchange in the former Ranch School stone primary-school building opened in September 1943. This branch, known as the Tech PX (Fig. 92), was near the entrance of the Technical Area and provided meals, fountain service, candies, magazines, tobacco, etc. In June 1944, another branch, known as the SED Exchange, was opened to serve the Special Engineer Detachment.

The Pee Wee PX, started in September 1945, handled soft drinks, and beer in packages, and sold cigarettes state-tax-free to Exchange employees and to civilians permanently employed and living on the Post. This PX also maintained a film-developing service to Santa Fe, which was a convenience to residents. It, further, took over the laundry-distribution station from the Trading Post.

An appliance store was established late in 1946, selling electrical equipment and appliances, as well as servicing radios and small electrical units for Project employees. A tailor shop was set up in one of the old ranch buildings.

In November 1946, the Post Exchange opened the Christmas PX. This store handled items suitable for Christmas gifts and included many things difficult to obtain in most cities, such as electrical and mechanical appliances, radios, and luggage. This operation was closed in mid-December.

Several further activities under Exchange jurisdiction were let on a concession basis. All concessionaire contracts were executed for one year, and provided that the Exchange would receive 10% of the gross receipts after deduction of state sales tax.

Fig. 90. The Trading Post was the original Ranch School Trading Post. The Service Club is to the left.

Fig. 91. Service Club, with beauty shop and barber shop at right.

Fig. 92. The Technical Post Exchange.

The garage and filling station (Fig. 93) was under such a contract. It was originally a direct PX operation, but a concession was let to Mr. J. W. Miller, in August 1945. The filling station building was a small warehouse, T-61, later enlarged by the addition of a four-stall garage, to care for the increased demand for gasoline, tire repair, car washing, greasing, and mechanical repairs. Through December 1946, this concession produced a total revenue of $11,846.62 for the Exchange.

Another concessionaire contract was made in July 1944, with Mr. C. K. Noel, for the cleaning and pressing shop. In October 1946, Joseph, Phillip, and Theodore Ferris became the new operators. Until this shop was set up, all dry cleaning and pressing was sent to Santa Fe and distributed on its return through Building T-8.

Another concession, a shoe repair shop, produced $45.93 for the Exchange between December 11 and 31, 1946. The following indicates the growth of the Exchange operation.

TOTAL CASH RECEIPTS

July 1943	Sept. 1943	Oct. 1943
$ 8,178.17	$ 22,520.64	$ 21,237.37

Feb. 1945	Feb. 1946	Dec. 1946
$92,568.32	$100,054.42	$192,510.12

Labor conditions were such that proper and adequate personnel could not be found to conduct many of the operations. WAC's worked in the Exchange office and as counter girls, and enlisted men worked as warehousemen, salesmen, and truck drivers. These positions were filled by selecting military personnel who had the proper qualifications. All worked full time on their Post Exchange assignments, being paid half their base pay in addition to the regular pay. Even when civilian personnel became available, it was still necessary to use military personnel because of

Fig. 93. The garage and filling station.

expansion. The total of 17 Exchange personnel increased to 43 full-time military, 117 part-time military, and 145 civilians by February 1946. By December 1946, the total was 167 civilians, 43 full-time military and 17 part-time military, showing a decided decrease of military and increase of civilians.

Security regulations prevented the Exchange from receiving the assistance normally granted by the Service Command. Consequently, the Exchange officer made all purchases direct from suppliers. No vendors were allowed to visit the Project, and considerable time was spent in locating vendors who could supply the amount and variety of merchandise necessary to take care of such an isolated post. Exchange items were sold to civilian and soldier alike. Civilian auditors made quarterly audits of the books, the first on May 25, 1944.

The Los Alamos Community Association

The Los Alamos Community Association was formed in 1946. Its principal functions were to handle, through concessionaire contracts, certain enterprises not included under the Post Exchange, and to use the proceeds to subsidize the library, athletic fields, a proposed swimming pool, and other activities for the good of the community. A five-member council was set up. They, in turn, hired a part-time manager to arrange for bids and administer the incoming funds. The roller skating rink was one concession established under this program. The contract on this concession was awarded to Elton J. Bonneville, in September 1946. The facilities consisted principally of a canvas tent and a wood floor, furnished by the concessionaire. A sliding scale of revenue went to the Community Fund from the receipts of this business, as follows.

5% on a weekly gross of up to $250.00
7½% on a weekly gross of up to $350.00
10% on a weekly gross of more than $350.00

Direct Concession Contracts. Concessions were sometimes set up directly between the War Department and the concessionaire. One such enterprise was the Los Alamos Drug Store. Bids were requested by the Operations Division, and applicants were carefully checked. It was felt that the successful bidder should be a qualified pharmacist, and should live on the Project to be available for night calls. Thurman E. Gunter was selected, and the store opened in November 1946, in an original ranch house. The contract was based on the same stipulations as the Post Exchange concession contracts; the War Department was to receive 10% of the gross receipts after deduction of the state sales tax.

Post Office

Sundt Construction Company maintained a fourth-class post office at the Project from early 1943 until November, with Ray Schoen as the first postmaster. The Post Exchange also handled mail deliveries through the General Store. This was not an authorized Post Office but merely a distribution point for Post residents. This operation, staffed by one WAC and three GI drivers, also sold stamps and Express Money Orders for the further convenience of the residents.

Mail for the Military units was separated in the Santa Fe Post Office and delivered directly to unit headquarters. All mail for the Technical Area was sent directly to the mail room in T Building, which was under the direction of Priscilla Greene. Project mail was brought up from Santa Fe by military drivers in an open weapons carrier. Three daily round trips were made regardless of the weather.

Mail service continued under the jurisdiction of the Post Exchange until Feburary 1945, when a postal officer was appointed to set up a new Post Office according to U.S. Engineer specifications. The new Post Office building was completed in May 1945. Money orders were still issued by the Post Exchange and remained their responsibility until October 1945. Then a Money Order Unit No. 2 was established under the Santa Fe Postmaster, and money orders and COD packages could be handled directly for the first time. Two Army mail clerks were in charge of these funds.

Police Department

Military Police. When construction began, civilian guards were assigned to internal security and remained until the first military personnel arrived to assume guard duty late in April 1943. The Military Police unit assigned to guard duty had been organized at Ft. Riley, Kansas, in March and designated as Provisional MP Detachment No. 1. The original strength was 196 men and 7 officers, all of whom met overseas requirements and volunteered for that type of assignment.

Soon after assuming the guard responsibility, the detachment was divided into four patrols of approximately 35 men each. Each patrol was on duty for 8 h and off duty for 24 h. The Sergeant in charge of each patrol acted as Commander of the Guard, and one officer was designated Officer of the Day. The Commander of the Guard was the only roving noncommissioned officer. This system was later revised to have an officer assigned to each patrol as Commander of the Guard. Privates were assigned as jeep drivers to roam the Site.

The unit had been organized with a mounted section, with the expectation that mounted sentries would be required. As many as six mounted posts were in effect from time to time, but they were found to be impractical and the last was abandoned in January 1944. After that, the 134 horses were used only for periodic patrols and recreational purposes. The number of horses was gradually reduced by disposal and transfer until none remained.

From May 1943, until the end of the year, half of the guards on duty were posted inside the Technical Area, guarding special buildings and the incomplete fence line. As construction neared completion, sentries were removed from inside the Technical Area, and two perimeter foot patrols plus gate guards replaced them. A fire guard with a time-clock system was on duty at night inside the Area.

Construction at new sites, beginning in July 1943, required additional MP personnel. On September 4, 1943, 99 additional men were assigned to the Detachment. The 8-h duty tour continued through the winter of 1943-1944, but proved too long for alert duty, and in April 1944, the guard personnel were reorganized into three patrols on a 24-h duty system. Privates were posted for 2 h on duty and 4 h off. Noncommissioned officers worked 6 h and rested 6 h for their 24-h period of duty. During the summer of 1944, 25 escorts were required to accompany uncleared workmen inside the Technical Area. This made it necessary to reorganize the guard into two patrols working alternate 24-h periods.

After fall of 1943, the patrols were subdivided into two sections, each under a section Sergeant equipped with a radio vehicle. The Commander of the Guard and his assistant, a Staff Sergeant, each had a radio vehicle. The Officer of the Day and the Commander of the Guard constituted the officer personnel.

With the growth of the Project, the authorized strength of the Detachment was increased until it reached 486 men and 9 officers. On December 31, 1946, the guard was manning a total of 44 full-time posts, requiring 115 men every 24 h. In addition to the regular guard, escorts for construction then in progress required the use of approximately five extra guards per day. Total strength of the MP Detachment approached 500.

All military personnel responsible to the local headquarters were not at Los Alamos. A unit of the MP Detachment was stationed at that part of the Alamogordo Bombing Range known as "Trinity." This unit, under the command of 1st Lieutenant Howard C. Bush, was in charge of security of the atomic bomb testing installations during the preparation for and the actual testing of the atomic bomb, on July 16, 1945.

The Military Police, under the supervision of the Provost Marshal, were the primary law enforcement officers of the Project. The Motor Patrol Section patrolled the roads, apprehended traffic violators, issued summonses for appearance before the court, and testified in resultant proceedings.

Community Law. Until July 1946, civil law enforcement was largely a matter of community cooperation. If the violator admitted to the offense, he was permitted to pay a stated fine, which was disposed of for the community through the Town Council. If he pled "not guilty," the individual was asked to appear before the Town Council where if found "guilty," he was asked to pay the same stated fine for his violation. The jurisdiction was granted voluntarily; in other words, whether or not one paid the fine was a voluntary action, and no legal measure could be brought to bear.

U.S. Commissioners Court. On July 7, 1946, necessary arrangements were made for trial of civil petty offenses before a United States Commissioner, Hon. Albert T. Gonzales, designated by U.S. District Judge Colen Neblett to try such cases. The court was established outside the Project boundary, to ensure that trials would be public. Authority to hold court at Los Alamos was granted with the knowledge and approval of Everett M. Grantham, U.S. Attorney for the District of New Mexico, to provide a means of punishing petty offenders. Petty offenses were defined as all those the penalty for which did not exceed confinement in a common jail, without hard labor, for a period of six months, or a fine of not more than $500.00, or both.

By way of statistics, the following figures were made available by Commissioner Gonzales and Capt. Harry Wise, Provost Marshal.

Offenders arraigned, July 10 - Dec. 31, 1946	712
Convictions on above with resulting fines	653
Acquitals or dismissals	59

Traffic violations constituted approximately 95% of all offenses; the others were such violations as simple assault, breach of peace, and disorderly conduct. More serious offenses were tried before the U.S. District Court Judge, but arraignment could be made before the Commissioner.

Fire Department

Until April 1943, fire protection was provided by the construction contractor. It consisted of one rented fire truck, equipped with a 150-gal booster tank and a few tools. Some 135 portable, pump-type, water-tank extinguishers were furnished to the construction contractor for distribution in completed buildings.

In March 1943, the Project received two Class-500 pumper fire trucks, each with standard equipment, including 2,000 ft of 2½-in. hose. In April, Fire Chief Edwin Brooks and six firefighters reported for duty, thus relieving the construction contractor. These firemen were all civilians and, with their families, were assigned to apartments in the area. The fire station had no quarters for personnel. In August, a Class-300 brush-fire fighting truck was acquired. In early October, the civilian firefighters, except the chief, were replaced by nine enlisted men. Quarters had been built at the Fire Station, and the change was made primarily to relieve the critical housing shortage, as well as to increase control of personnel.

From April through December 1943, the Fire Department answered 27 alarms as follows.

Building fires	8
Brush fires	5
Asphalt-kettle fire	1
Private vehicle fires	3
Furnace calls, no fires	8
Standby for trash burning	1
Demonstration	1

Estimated fire damage to Government property was $380; to private property, $765.

During 1944, personnel were increased to 14 enlisted men. In June, Fire Station No. 2, located near Anchor Ranch, was commissioned. It was manned by one sergeant and four enlisted men and was given the Class-300 brush truck from Station No. 1. Later a 3/4-ton weapons carrier equipped with back-pack water fire extinguishers and forest-fire-fighting tools was added. However, the brush truck had only two-wheel drive, so the nature of the terrain restricted its usefulness. A trade was arranged in November with Bruns General Hospital in Santa Fe, for a truck of the same classification but with four-wheel drive.

In December, a Class-325 brush-fire-fighting truck with four-wheel drive was acquired for Station 2, to handle an increasing number of brush and forest fires caused by operations at the site. Particularly because of the combustible construction at the town site, there was an appreciable possibility of having two or more alarms at one time, and a Class-750, pumper-type, fire-fighting truck was acquired in September for protection against this hazard.

During 1944, Station No. 1 answered 61 alarms, as follows.

Building fires	17
Hay fire	1
Brush fires on Post	3
Brush fires at Sites	5
Private vehicle fires	2
Government vehicle fires	1
Lumber-shed fires	2
Furnace calls--no fires	25
Standby, at sites	1
Covering Station No. 2	1
Oil-tank fires (one false)	2
Practice alert	1

Estimated damage to Government property was $4,253; to private property, $9,000. Of the latter,

$8,875 was from the loss of a truck and its load of electrical equipment about five miles from the town site.

During 1944, Station No. 2 answered 65 alarms, as follows.

Building fires	1
Trash fires	1
Forest and brush fires	39
Pumping out flooded steam pits	6
Investigations after fires	9
Standby at site operations	9

Estimated damage to Government or private property was nil. In addition to fire fighting, Station No. 2 delivered 154,265 gal of water to the various sites.

In May and June 1944, under supervision of the Forest Service, the Fire Department controlled the burning of about 10 square miles of brush and forest covering the area used and to be used at the outlying sites. This operation necessitated 24-h patrols, supplied mainly by the Military Police Detachment.

During 1945, the Fire Department was enlarged to include a civilian Chief, 3 military Deputy Chiefs, and 43 military firemen. The equipment consisted of the two Fire Stations and 12 pieces of mobile apparatus: In Station No. 1, a 750-gal/min triple combination engine, three 500-gal/min coordination pumping engines, and one 300-gal/min brush truck; and in Station No. 2, one 500-gal/min triple combination pumping engine, five brush trucks, and one weapon's carrier.

During 1945 there were 379 fire calls for both stations.

	Station No. 1	Station No. 2
Chimney fires	12	0
False alarms	17	1
Wooden shed fires	4	1
Brush fires	14	172
Hot-furnace calls	55	0
Garage fires	2	0
Rubbish fires	14	19
Sawdust-pile fire	1	1
Unnecessary alarms	14	2
Oil-truck fires	1	1
Electrical fires	2	6
Explosions	1	0
Defective alarm system	6	0
Oil-stove fires	3	0
Trailer fires	4	0
Structure fires	15	6
Vehicle fires	4	1
	169	210

"Unnecessary alarms" were those turned in when people thought it necessary to call the Fire Department but no fire existed. "False alarms" were those turned in deliberately and often maliciously by people who were well aware that no fire existed. These were arbitrary terms used by the New York Fire College, Long Island City, N.Y. The 1945 estimated fire damage to Government property was $148,790, of which $124,700 was C Shop (Fig. 94). Private property damage was estimated at $4,500.

In August 1945, a large new central fire station, known as Station No. 2, was completed and put into operation. This new structure (Fig. 95) replaced the old Station No. 1. A practice tower and a hose-drying tower were erected adjacent to this structure. The equipment and apparatus attached to old Station No. 1 was shifted to the new operation, and two new 500-gal/min pumps and two 750-gal/min pumps were acquired, with 5,000 ft of 2½-in. hose, 3,000 ft of 1½-in. hose, new fog nozzles, six 300-gal/min brush trucks (4 x 4), one 300-gal/min 2-wheel-drive brush truck, and 5000 ft of booster hose.

By December 1946, personnel had increased to 1 chief, 2 assistants, and 90 military firemen. The apparatus and equipment consisted of the above mentioned additions, as well as:

3	750-gal/min pumping engines
4	500-gal/min pumping engines
12	300-gal/min brush trucks
1	300-gal/min weapons carrier
4	700-gal/min trucks
1	750-gal/min crash truck

The department also had six mobile radio cars and two fixed stations, one in each fire house.

There were 411 fire calls during 1946, as follows.

Fig. 94. Technical Building C, (rear view).

Fig. 95. Central Fire Station and tower used for practice jumps and hose drying. The power house is visible at the right. Buildings in center and left background are property offices and warehouses.

	Station No. 1	Station No. 2
Chimney fires	1	0
False alarms	46	0
Brush fires	42	179
Defective heating equipment	16	1
Vehicles	15	2
Rubbish fires	6	0
Unnecessary alarms	4	0
Unknown	5	2
Electrical	9	0
Overheated equipment	7	2
Asphalt spreader	5	0
Sparks on roof	5	0
Sawdust	3	0
Cigarettes	10	0
Classified material	5	0
False sprinkler supervisory	39	0
Miscellaneous	3	0
Standby alarms	4	0
	225	186

All community structures were equipped with portable extinguishers, totaling as follows.

Water extinguishers	1811
Soda-acid extinguishers	1000
Foam extinguishers	996
Carbon dioxide extinguishers	1800
Carbon tetrachloride extinguishers	1200
	6807

In April 1946, a Fire Prevention Section was organized, employing 25 persons whose duty it was to check fire hazards and fire equipment, maintain fire extinguishers, and educate employees in fire prevention.

Schools

During the summer of 1943, it became apparent that school facilities would have to be provided for the children of the resident families. There had been a small elementary school at the Los Alamos Ranch School, but the building was too small. Dr. Walter W. Cook, Professor of Education, University of Minnesota, was called into consultation with the Director and Division Leaders of the Project. These men mapped out the structures needed and the policies to be adopted for a school system that proved adequate and satisfactory.

The original building (Fig. 96) was designed by Cook and Brazier. Emphasis was placed upon simplicity and efficiency of classroom surroundings, proper lighting, and up-to-date apparatus. The first building had four elementary-school rooms and four high-school rooms (Fig. 97), ample for the enrollment of 140 children. However, additions were required every year. At the end of the 1946 school year, there were over 350 students enrolled. In the fall of 1946, another wing of five rooms was completed to house a new library, home economics class, and general science classes.

Adequate salaries attracted excellent teachers to this school. The average weekly salary was $72.57. Many of the teachers were wives of employees at the Project. There were nine instructors on the original faculty for both elementary and high school. By the end of 1946, there were 18 grade-school teachers, 13 high-school instructors, and a superintendent for the somewhat more than 350 students. This favorable ratio of children to teachers permitted unusual individual assistance and instruction and gave the students a wide choice of subjects.

The school operated as a free, public institution. The Government paid expenses, with the University of California originally acting as the Government paymaster, paying all school salaries and handling all requisitions for textbooks and school equipment. In June 1946, the Zia Company took over this function.

The regular school system was supplemented by a nursery school (Fig. 98) for children from two to five years old. Its primary purpose was to care for the younger children of mothers employed on the Project. A director, four teachers, cook, maid, and janitor composed the staff.

Medical Service

In accordance with instructions from the Laboratory Director, the University of California hired Dr. Louis H. Hempelmann and Dr. James F. Nolan in the latter part of February 1943, and shortly thereafter, three civilian nurses, Sarah Dawson, Harriett Petersen, and Margaret Schoppe, to care for the general health and welfare of the personnel at Project Y. The

Fig. 96. The first school building. The Jemez Mountains are in the background.

Fig. 97. A class in Los Alamos High School.

Fig. 98. Nursery school and play yard for preschool-age children.

preliminary arrangements were made orally through personal conference among Dr. Oppenheimer, the Laboratory Director, and Doctor Hempelmann and Doctor Nolan.

The physicians' duties included protecting personnel from industrial hazards and providing general medical care for Project residents. Specifically, the two physicians were hired with the first duty in mind, as both were trained along these lines. Dr. Hempelmann had trained in internal medicine and radiology; Dr. Nolan, in obstetrics, gynecology, and radiology. As a general plan, Dr. Hempelmann was in charge of industrial hazards and Dr. Nolan was in charge of general health, and each assisted the other.

It was at first thought that both doctors would be commissioned in the Army, but this procedure was changed because of changes in the nature of their work. Dr. Nolan was commissioned a First Lieutenant in June 1943, and, as Project Surgeon, was directly responsible to the Commanding Officer of the Project, in military channels, for medical care of personnel and maintenance and supply of the medical facility. His military rank also permitted more effective liaison between Project personnel and Bruns General Hospital in Santa Fe. Dr. Hempelmann was soon fully occupied in consultation in the scientific program and could not assist in the practice of medicine.

The doctors arrived in Santa Fe in March 1943, at which time the Site was not yet inhabited. During early April the two doctors and three nurses assumed their responsibilities in a five-bed industrial-type infirmary. Full equipment was not available until several months later. Although the Project was supervised by the Army, the medical facilities were originally considered a University of California function. In late February 1943, the Surgeon General arranged with Bruns General Hospital that all civilians and their dependents residing at the Project could receive full medical and dental care from this source. Bruns General Hospital was located about 50 miles from the Site over what was then a rather tortuous road. Despite this arrangement, care of Project residents at Bruns or by private physicians was discouraged for security reasons. Residents were encouraged to seek medical aid at the site, without charge, in order to prevent traveling outside the area.

Military personnel were originally cared for by 1st Lt. J. J. Horowitz and a staff of seven enlisted men. A three-bed Army infirmary was available at Los Alamos, for use of all military personnel. Cases severe enough to warrant hospitalization were sent to Bruns General Hospital in accordance with Army procedure.

The original plan of hospital procedure was, in the main, an oral and tacit arrangement. Subsequently, there were changes in the technical program and also in security arrangements. These necessitated changes in the medical functions. The staff was soon found to be much too small for the increasing population. A pediatrician, 1st Lt. H. L. Barnett, was added to the staff in July 1943. In the beginning, hospital maintenance was done by two untrained Indian boys, who were found to be incompetent, and, because civilian medical orderlies could not be obtained, eight medical enlisted men were borrowed from the detachment working under Lt. Horowitz. After that, medical orderlies and technicians were supplied from Army personnel.

In the fall of 1943, Capt. Horowitz was transferred; and Lt. Paul O. Hageman, a specialist in internal medicine, replaced him in January 1944. In June 1944, action was initiated and approved to expand the larger infirmary to Station Hospital classification. (See letter and exerpt from memo that follow.) This was done in the fall of 1944. With this arrangement, the military infirmary was still used for the Army sick call, but, because of the lack of nursing supervision, Army personnel were hospitalized at the larger installation (Fig. 99).

Lieutenants J. E. Brooks and A. M. Large, specialists in nose and throat and general surgery, respectively, were procured during the latter part of 1944, to provide a medical staff with specialists in each of the five major fields. An administrative officer, 2nd Lt. C. M. Berg, also was added at this time. From the middle of 1945 through 1946, the staff of doctors, nurses, and technicians increased materially. Mrs. Darol Froman was the first pharmacist, but she worked only part time. A full-time pharmacist and an x-ray technician were attached to the group in May 1944, and a dietician in December 1945.

Medical care was provided for all Project residents. Emergency treatment was given all nonresidents who had accidents on the Site. The construction contractors and their subcontractors were charged for emergency medical treatment

Fig. 99. The hospital.

This document consists of 2
pages. No. 4 of 7 copies, Ser.A

ARMY SERVICE FORCES
United States Engineer Office
Manhattan District
Oak Ridge, Tennessee
P.O. Box E

In reply
refer to MD 632

22 June 1944

Subject: Hospital Requirements at "Y"

MEMORANDUM TO: Major General L. R. Groves

1. The steady increase in population during the past three months and the probable increase of 25% within the next year have created a problem in hospitalization.

2. Recent inspection and study of the hospital and clinic facilities at "Y" indicate that they have become dangerously inadequate within the last month.

a. The hospital at present is designed for nineteen beds. However, twenty-four have been crowded in, and are fully occupied most of the time.

b. The military infirmary was designed for three bed patients, but ten beds have been crowded in and are fully occupied at frequent intervals.

c. There is no flexibility in the present facilities to meet the fluctuating demands of the various contingencies as well as emergencies which may arise in any group of this size and activity.

3. The future hospital and military infirmary requirements are difficult to estimate because of various factors, some of which are obscure. Therefore, the present recommendations are based on local experience of the past year plus certain obvious safe-guards required by this unusual type of community geographically isolated from the usual facilities of a city.

a. The large percentage of "intellectuals", both old and young, in a civilian group who, when subjected to the strain of transplantation and high-pressure work, are more susceptible to mild medical ailments than usual, and will both require and seek more medical care than the average person.

b. The high proportion of "4-F's" require medical supervision and care.

c. Over one-third of the civilians are single and must be hospitalized for even mild illness because they cannot be cared for in the dormitories.

A TRUE COPY:

Edith C. Truslow

EDITH C. TRUSLOW
2d Lieut. TC

Subject: Hospital Requirements at "Y"

d. Approximately one-fifth of the married women are now in some stage of pregnancy. (The birth rate over the nation elsewhere is decreasing.)

e. Approximately one-sixth of the population are children, one-third of whom are under two years of age.

f. Due to lack of house-hold help, the ill patient must be hospitalized longer, particularly if it is the wife who is ill.

g. Approximately one-half of the population is military, one-third female, and two-thirds male. According to regulations, when ill enough to be off duty, even though convalescing, they must be hospitalized.

3. It is recommended that the present hospital and appropriate facilities and personnel be enlarged to a fifty to sixty "Sick patient" capacity, depending upon the most efficient number for the basic plumbing and other services, with the distribution approximately as follows:

a. Eight single rooms in an isolation unit for infectious disease.

b. Sixteen cribs for pediatrics.

c. Approximately fourteen single rooms for the very ill cases, post-operative and obstetrical patients.

d. The remainder distributed in two to six bed units as convenient to services and space requirements.

4. It is recommended that sufficient open-spaced solaria accommodating twenty cots or beds, and one or more large screened porches for twenty cots be made available as a minimum for emergency and epidemic situations, and convalescent civilian and military personnel as required.

5. It is recommended that additions be made to the present hospital rather than starting new construction at another site.

STAFFORD L. WARREN,
Colonel, Medical Corps

cc: Col. Nichols
Lt. Col. Ashbridge

A TRUE COPY:

-2-

Edith C. Truslow

EDITH C. TRUSLOW
2d Lt. TC

Excerpt from June 9,1944, Memo to Lt. Col. Ashbridge, from Capt. James F. Nolan.

> *There are now available in extremely crowded conditions, only 24 beds. The situation is well illustrated by the following: there are three beds now assigned to obstetrical cases and at the present time there are five obstetrical patients hospitalized. There are ten pediatrics beds and on the basis of use, sixteen would be required to meet the demand, exclusive of infectious disease. There are four surgical beds. These are full, and four times as many should be available as a minimum. There are seven beds assigned to medicine -- all of which are also full -- and a little over twice that number should be available. Experience has shown that hospital facilities of only 24 beds are not only inadequate, but dangerously so; and that approximately 55 - 60 beds will be required to handle the present population with any degree of safety.*

according to established rates on file at the hospital. All civilian in-patients were charged one dollar per day subsistence. All collections were made through the Finance Section, including the subsistence charge and the fees charged to the contractors.

In June 1946, the hospital operating procedure was changed to a cost-plus-fixed-fee contract basis under the Zia Company. Charges for non-job incurred accidents and illness were initiated at this time. At the end of December 1946, the total hospital bed capacity was 107.

The sanitary aspects of the public health functions were under Army jurisdiction, with the Project Surgeon acting as advisor. Functions under this system included water supply, sewage and garbage disposal, and fly and other pest control. Mess-hall inspections and routine examination of food handlers were carried out by the medical staff. Also, the population was immunized against typhoid, tetanus, and small pox. A school program included daily examination of children, immunizations, and health and first aid lectures.

Initially, communicable diseases were not reported, but later they were reported through military channels to Oak Ridge. However, if individuals with venereal or other communicable diseases left the Project without treatment, they were reported directly to the Director of Public Health in Santa Fe.

Dental Service

The original policy was to have a dental officer from Bruns General Hospital come to Los Alamos two or three days a week to care for emergency and regular appointments. A regular dental office had been built in the hospital here, equipped with three units. Lt. Richard Mosgrove was the first officer from Bruns. Later, Lt. Gerald Bigelson from Bruns was offered permanent assignment here and accepted in March 1944. He was the only dentist until December 1944, when Capt. Nathan Peretzman came. The staff was increased to three in September 1945, with the addition of Capt. Belgorod. Dental care was given not only to all military and civilians at Los Alamos, but also to those from Sandia who required it. At first, these cases came up from Sandia, but as that installation grew, one of the dentists made several weekly trips there to handle appointments. The equipment and supplies were all standard Army type. Requisitions were handled through the Project Medical Supply and, in turn, procurements were made from Medical Depots. There was no difficulty in getting supplies, but there was delay, as shipments were often not made direct because of security. The staff in December 1946 was made up of four civilian doctors, three of whom had been at Los Alamos in the Army.

Veterinary Service

The Veterinary department was organized in April 1943 as part of the MP Detachment. Its original duty was to care for all animals (as many as 134 horses and 15 war dogs were used at one time, for guard duty, and there were numerous pets and mounts owned by residents), and to inspect all foods of animal origin, including milk. Lt. R. E. Thompsett and Sgt. James Mattox comprised the first staff, which increased in March 1944 to the one officer and three enlisted men, all qualified in veterinary work. Their functions included care of army mounts, dairy inspection throughout the area to ensure proper standards for milk, candling eggs, daily inspection of meats, control of contagious diseases in small animals, care of animals used in experimental work, treatment of civilian-owned animals, and reporting and administration.

Supply and Procurement

Supply. The first property was acquired when the Manhattan Engineer District formally accepted the Ranch School in February 1943. At that time, an Accountable Property Officer was set up in New York, and the Los Alamos Account was carried on Memorandum Receipt from Oak Ridge. This was only for three months, however; Col. J. M. Harman was appointed Accountable Property Officer in May 1943. He set up a Division with an Accountable Property Section and a Property Receiving Section. Here, too, the unexpected and unanticipated expansion in volume of receipts greatly hampered the section, as did the scarcity of competent personnel. Also, the account was under the rigid restrictions imposed by Security. Consequently, normal property-control measures became somewhat ineffective, making it necessary to inaugurate an extension to the usual measures to include the wide-flung activities.

The functions of the Post Supply Office included the following services to the Sant Fe Area Site, the Santa Fe Office, Trinity Site, and Sandia Base.

Furnishing building equipment and maintenance supplies for both the Technical Area and the Post.

Supplying military detachments.

Furnishing housing equipment.

Furnishing office equipment and supplies.

Furnishing motor vehicles, automotive parts, and supplies.

Meeting normal Post requirements.

Furnishing construction materials unobtainable by construction contractors.

There was a certain amount of difficulty in setting up a Standard Operating Procedure, because of changes in applicability of different directives. The account operated on Chapter IX, Orders and Regulations, Corps of Engineers, until May 1943, because no directive had been issued to cover the operation. In late August, Circular Letter 2398, from the Office of the Chief of Engineers, was made effective, and all items with a monetary value of $10.00 or less, were to be classified as expendable and accounted for as such. After another change instituted early in 1946, the account was operated strictly in accordance with War Department Supply Regulations. There was no complete property audit from the beginning of the Project until the Spring of 1946, when the heavy pressure of the work had been relieved and there was time for such necessary checks.

Procurement. Procurement was also the responsibility of the Accountable Property Officer. Colonel Harman was in charge of Procurements until July 1943. The office was originally at the Project, but in mid-1943, it was moved to the Bishop Building in Santa Fe, with Mr. Peter A. Curran in charge. This section's function was to review all requisitions from various Divisions. If these requests were in order, the material was sought first through Army channels and next through commercial jobbers by Purchase Order. Besides being the liaison point with these commercial accounts, Procurement expedited shipments and arranged transportation.

It was understood that the University of California Procurement Section would deal solely in highly technical and scientific supplies. The Post Supply Officer provided all supplies considered nonscientific for the Technical Area. At times Post Procurement received technical shipments which were sent immediately to the Technical Area.

Fuels procured included coal, diesel oil, distillate, and kerosene. Except for negligible quantities purchased at the beginning of activities to meet initial requirements, coal consumed at the Project was procured through established Quartermaster Depots; that is, contracts were executed by the Quartermaster to cover varied requirements. No unusual problems arose in providing adequate coal for domestic and technical use, as local mines produced coal meeting detailed specifications. Large quantities of diesel fuel were procured for operation of the power plant. All such procurement was made under appropriate Treasury Procurement Schedules in accordance with existing rules and regulations. Substantial quantities of distillate and kerosene were procured. Equipment burning these fuels proved far more efficient than similar coal-fired heating equipment, so more of these fuels were purchased in successive years. In early 1945, it became a policy to purchase oil-burning rather than coal-burning equipment for domestic and technical use, partly because of longer hauls involved in obtaining necessary coal.

Intelligence and Security.

The Project's Intelligence and Security Office began operation in April 1943, with the arrival of the first Intelligence Officer, 2nd Lt. Curtis Clark. At that time the office had only two military members, but, with Project growth and the expanded Security program, it was gradually enlarged to 28 military members and 7 civilian employees in March 1945, 115 military and 18 civilians in April 1946, and 84 military and 48 civilians in December 1946.

Project Security included operation of the pass system, visitor control, and the guard system, including shipments, personnel, materials, and mechanical guard installations. The pass system was designed to prevent unauthorized entry to the Project. It was simple in operation to avoid confusion, but effective as to purpose. To facilitate the flow of materials and construction workers, construction contractors and designated employees of the U.S. Engineer Warehouse in Santa Fe were authorized to sign carrier and construction employee passes. To expedite the admission of new arrivals to the Project, a pass office was established in Santa Fe, and was also used as the Project bus terminal. Until the end of the war, the only persons permitted to visit the site were those who came on Project business, whose presence at the site was necessary to the conduct of such business, and for whom permission was obtained from the Director or Personnel Director after authorization was first obtained from the Security Office. Later, friends and relatives were permitted social visits. Technical employees of the Project were permitted to visit other Projects, laboratories, universities, and scientific personnel, but only on authority of the Director and after clearance for the visit had been obtained from the Security Office.

The principal aim of the guard system, as executed by the MP Detachment, was to provide the greatest security with the least manpower, correlated to expedite rather than retard the work of the Project. Main guard posts were established at the two entrances to the reservation and the three entrances to the Technical Area, with others at the outlying sites. As an additional security measure, the entire Technical Area was enclosed with a mesh-wire fence, on which an automatic alarm system was installed.

Internally, the pass system was an integral part of the guard system. Passes were classified as "Project" or "Technical" area passes, and appropriate badges were issued in addition to the permanent pass, and worn at all times in the respective areas. Badges of one color and design were issued to staff members, group leaders, and other key personnel in the Technical Area, and to key military personnel, entitling them to admission to certain areas. Badges of another color were issued to non-key members of the technical group, construction employees in the Technical Area, and clerical employees.

As the work progressed, various test sites were constructed, each being operated by a different group within the Technical Area. A visitor list was set up for each site, and only those appearing by name and badge number on the visitor list were permitted to enter the site. Guards for classified shipments from the Project were supplied by the Security Office, as were couriers for documents, convoy guards, etc. All arrangements for meeting, delivering, and safeguarding incoming shipments were handled by the Security Office.

For the protection of certain key military and civilian personnel, armed guards were furnished by the Security Office. Members of that group were accompanied by an armed representative of the Security Office on all trips away from the Project. To conceal the identity of certain key personnel, a system of code names and numbers was set up for their use. Automobile registrations, drivers' licenses, income-tax returns, insurance policies, and food and gasoline rations for these people were handled by the Security Office by the use of code names and numbers, with no disclosure of their true identities. Their mail was sent and received by an indirect route to conceal their actual location.

All technical and scientific operations involving use of special materials were under the direct supervision of the Security Office. A special detachment for the purpose was composed of military and civilian guards. These special materials were placed under continuous armed guard because their loss would be enormously expensive and delaying.

Security installations included a 9½-ft-high woven-wire fence, with two barbed-wire top strands, surrounding the Technical Area, which was, in turn, enclosed in a similar fence surrounding the community structures also. A few similar fences surrounded outlying sites. Peep-proof board fences surrounded the Anchor Ranch gun site and V site. Some hundred and fifty nine 1500-W flood lights were mounted to light the vicinity of theTechnical Area. In addition to these precautions, a prowler system was installed on the fence surrounding the Technical Area and on another small fence surrounding building T-412. Unusual mechanical vibrations in the fence caused by an intruder produced an audible signal in a loud-speaker, and a visual signal in a circuit light, either of which could attract the attention of the sentry on duty. The system was not satisfactory, as the least vibration, such as a strong wind, set up the signals and caused more confusion and trouble than good. Consequently, it was not used after January 1945 and was finally removed entirely in May.

The Intelligence Section was primarily concerned with safeguarding classified information and investigating prospective employees. It was always a basic Project policy that each employee should have access to all the information required to do a competent and expeditious job. This created the problem of educating each to keep silent about his work and the Project. Each employee was given a detailed security lecture by an Intelligence Officer, in which he was informed that all information about the Project was classified and its dissemination was an offense punishable under the provisions of the Espionage Act. Each terminating employee and all members of the Special Engineer Detachment were given similar lectures.

All Project personnel, both military and civilian, were cleared by the Security and Intelligence Office before assignment or employment, a procedure often requiring much time. This clearance included both administrative and G-2 approval. Employees could be transferred to the Project from other Manhatten District projects only by authority of the Commanding General upon petition by the Director.

Because restrictions on other phases of personal freedom at the Project were rigid, the rumor was started (based, perhaps, on the evidence that a few letters to individuals had been opened by error) that unannounced censorship was in effect. This rumor grew to such proportions that General Groves ordered an investigation which, of course, indicated that no action of this kind had been taken. However, because the issue had arisen, it was deemed advisable in further safeguarding classified information to institute censorship. Thus the rumored censorship became actuality when Dr. Oppenheimer approved the measure. Censorship of all mail was instituted in December 1943 and ceased shortly after the end of the war. This function was organized under the provisions of War Department Training Circular No. 15, which provided that:

> *"When the military authorities deem it necessary in the interest of security, military censorship may be effected over all communications entering, leaving, and within any area, or to or from any personnel, under military jurisdiction within the continental limits of the United States."*

Censorship was conducted at a point outside the Project limits, by trained censorship officers, in strict accordance with Army Regulations. Only official mail was exempt from censorship, and all Project personnel, military and civilian, were advised that personal communications were being censored. For the same purpose, limited monitoring of long-distance telephone calls was conducted under the supervision of the Security Office, and all incoming and outgoing telegrams and teletype messages were reviewed.

The Security Office gave instructions on the classification of documents and the handling of these documents and materials. A system of code words was instituted for use in this connection. Vacations and pleasure trips by Project personnel could be taken only on authority of the Security Office. Project employees' families were not permitted to live in Santa Fe, or in any other town within a radius of 40 miles of the Project, nor could Project employees receive visitors.

Liaison With Other Agencies

Relations with outside organizations, primarily the State of New Mexico, were always extremely cordial. This was very important because, to ensure security, there were many unusual requests relative to compliance with State laws, such as obtaining drivers' and automobile licenses and paying State income tax. Because the names of many University of California employees could not be disclosed, the Commanding Officer agreed with the State that information would be available in Project Headquarters whenever it was needed, and the State thereupon issued licenses and accepted income-tax forms on the basis of numbers in place of names. When Ration Book No. 3 was issued, in the fall of 1943, special arrangements were made with the Rationing Board of Sandoval County, whereby the books were issued at the Project and the applications were kept on file at headquarters, to be available only should the County of Sandoval need to refer to them.

An attempt was made to arrange for Project residents to vote in the 1944 Presidential election, but it was unsuccessful because the personnel lived on a Federal Reservation, and only those who had retained residence in other states and were able to secure absentee ballots could vote. This was unfortunate, especially because the State of New Mexico, in accordance with the Buck Act, nevertheless expected all employees to pay State income taxes.

Morale

Religion. Religious services were arranged shortly after the Project started. The influence of the early Spanish occupation is seen in the strength of the Roman Catholic Church in New Mexico. The Roman Catholic Cathedral in Santa Fe arranged to send a priest to the Project for Sunday Mass. Other services, including Mormon, Baptist, and Methodist, were held in various homes. In the spring of 1944, the Ministerial Association of Santa Fe (Protestant Churches) arranged to provide clergymen for services at the Project. These clergymen took turns and visited the Project as their own parish schedules permitted. This was not satisfactory because these services were scheduled on very short notice and during irregular hours. There was an obvious need for a regular clergyman who could provide his full services. An Army Chaplain, Capt. Mathew Imrie, a Protestant, reported for duty in August 1944. He immediately scheduled Protestant services and began a study of community needs. In the next month, Protestant Sunday Schools and Hebrew Sabbath Services were initiated. The three religious groups, Roman Catholic, Protestant, and Hebrew were given regular opportunities for formal worship. The Chaplain was given an office in the Big House, and was available for consultation. Major Imrie was succeeded by Capt. Kenneth Ames in August 1946.

The Chaplain's Office established the Choir and Glee Club, Church Council, Ushers' Group, Women's Church Guild, and Altar Guild. Calls and visits were made within the Project and to Bruns General Hospital. The Chaplain's community activities included work with the Boy Scouts, Girl Scouts, Parent-Teachers Association, National War Fund, Annual T.B. Christmas seals sale, athletic programs, and assistance to the Red Cross Field Director. He was also chairman of the Youth Council, whose seven members managed the Youth Center, Youth Library, and playground.

Military Educational Facilities. As Post Information and Education Officer, the Chaplain advised the Military Detachment Commanders and lectured to enlisted personnel on various subjects. The United States Armed Forces Institute courses, texts, and material were charged to the Chaplain's office until September 1946, then shifted to the Special Services Officer. Under this program, high school and college level tests were given, and many of the military received high school diplomas, or certificates of satisfactory scores for presentation to universities. Regular night-school classes were held for four 3-h periods per week. Student enrollment reached a peak of 250, requiring 15 instructors.

Recreation. Organized recreation was started in February 1943, under the direction of Jerry Pepper, a former teacher at the Ranch School. The Ranch School facilities continued to serve: five tennis courts, a football and baseball field, a ski run, and an ice skating pond. Two outlying campsites, Camps Hamilton and May, were transferred with the school facilities and were available for camping trips.

In November 1943, recreation became an Army function under Special Services. Under the new regime, four additional tennis courts, four softball fields, a makeshift golf course (since abandoned and replaced by an 18-hole course), and a bowling alley were added.

The residents themselves organized many groups, including: Women's Club, Boy Scouts, Cub Scouts, Girl Scouts, Brownies, Spanish Club, Home Arts Club, Dramatic Club, Mountain League Baseball, Arts and Crafts Club, Sketching Club, Chess Club, Singing Group, Basketball Leagues, Ski Club, Community Welfare Association, Zia Employees' Association, Mesa Club, Officers Lounge, NCO Club, Veterans Club, Little Theater Group, Camera Club, Pottery Group, Victory Garden Group, Women's Softball League, Civilian Air Patrol, Fencing Group, Rifle and Pistol Club, Bowling Leagues, and Touch Football League. In addition to these activities, dance bands and dances, art shows, tennis tournaments, golf tournaments, monthly musicals, bingo games, Christmas pageants, track and field meets, and lectures by local authorities were scheduled. USO shows were not given for military entertainment because of security regulations.

Two theaters supplemented the recreational facilities. Theater No. 2 (Fig. 100), in the central area, had a basketball court, and was used for a school physical-education building. It had a screen and stage for such activities as movies, plays, dances, concerts, conferences and addresses. It was the meeting place for the Colloquia, Tuesday night conferences for all University of California staff members. The smaller Theater No. 1 was used almost exclusively for movies.

A small radio station (Fig. 101) was started in December 1946 to give extra entertainment to Los Alamos residents. Because of security restrictions, it was limited in power and could not be heard beyond the Project boundaries. The programs consisted mostly of records, Los Alamos talent, news reports, and lectures.

Transportation

A Motor Transportation Pool and Shop were established in July 1943. The pool at that time consisted of about 45 vehicles; until then each vehicle had been assigned to a specific person or

Fig. 100. Theater No. 2.

Fig. 101. A Ranch School building first used for substandard housing and later for the tailor shop. The temporary building at the right was originally a workmans' barracks, later converted into the Telegraph Office headquarters, radio station, PX Office, and newspaper office. The snowcapped mountains are the Sangre de Cristos.

group. A few vehicles remained assigned to the University of California and were not subject to pool control. Two WAC's and seven enlisted men drivers were originally assigned to make personnel trips and haul truck freight from railheads and airports at Santa Fe, Lamy, and Albuquerque. The following shows Motor Pool growth by personnel assigned.

	WAC	EM	Civilian
July 1943	2	7	
Dec. 1943	5	15	
July 1944	7	22	
Dec. 1944	9	30	
Mar. 1945	11	42	
Mar. 1946	5	34	31
Dec. 1946 (Zia Company)	---	---	116

The Motor Pool transported personnel to and from the Site, and transported all freight from railheads and airports. The operation increased proportionately with the growth of the Project. Other duties were to dispatch all vehicles necessary for Project construction and administration. Vehicles assigned to the University of California were under its control. These vehicles were dispatched like those in the central pool, and they were inspected and maintained there. The following indicates the vehicular growth of the motor pool.

	Vehicles
July 1943	45
Dec. 1943	250
Sept. 1944	400
March 1945	650
March 1946	1,181
Dec. 1946	1,180

The types of vehicles were as follows.

Sedans	177
Jeeps	125
Pickups	165

Carryalls	47
Weapons Carriers	171
Panel Trucks	12
Fifteen-Passenger Conv. Sedans	6
Armored Car	1
Buses	85
Trailers	78
Ambulances	13
Miscellaneous Trucks	279
Wreckers	9
Farm Tractors	2
Fork-Lift Trucks	7
Cranes	3

Some 249 vehicles were assigned to the University of California, the rest to military detachments, the hospital, maintenance crews, and construction crews.

Until July 1943, all freight required at the Site was hauled by commercial truckers at prevailing hourly scales. Security and economic reasons induced the project to try to haul all freight using Army trucks and soldier drivers. All materials and supplies were hauled initially from Santa Fe, partially over rough, steep, unpaved mountain roads. A round trip to the railhead at Santa Fe required 92 miles of travel. Because Albuquerque was the only major marketing center within a reasonable distance, it was often necessary to dispatch trucks to that point for essential supplies and equipment. Air Freight shipments received via ATC arrived at Kirtland Field, Albuquerque. A round trip to Albuquerque involved 228 miles of travel, and the wear and tear to mechanical equipment from such lengthy trips was high. In early summer of 1944, the entire road between Santa Fe and the Project was surfaced, thereby reducing this abnormal wear and tear.

For the first six months of the Project, two seven-passenger station wagons were adequate to haul passengers to and from Santa Fe. By 1946, approximately eleven 40-passenger buses were making the same round trip daily. In March 1945 and March 1946 there were approximately 30 busses in daily service to numerous points within 50 miles of the Project. By December 1946, approximately 518 passengers in 43 buses were making the daily round trip. There was no competition with commercial bus lines because the routes and schedules did not coincide with any established line. Operation of these buses was considered necessary in the interest of obtaining essential man power, because housing facilities were so limited. Such bus service was provided free to laborers, mechanics, maids, etc. because it was the only way adequate help could be obtained.

About 60 Indian girls used the buses each day, incident to their employment as maids by individuals housed on the Project. The maids were paid by the individuals utilizing their services. However, maid service was considered essential because there was a strong suggestion that wives or adult dependents should be employed while living on the Post. Furthermore, it was imperative to conserve the time of all scientific employees.

In April 1946, the Zia Company took over transportation, absorbing the buses, most of the heavy equipment, and the responsibility of conveying personnel and freight. The bus schedules were maintained, with 43 buses making round trips to nearby towns, three buses for school children, and seven other buses operating on the Project. Zia also installed a taxi service for official transportation only, following substantially the same procedure as that under which the Motor Pool had been operated.

The Motor Repair Shop, established in July 1943, was initially staffed with six soldier mechanics. Until then, all necessary repairs were made by commercial garages in Santa Fe. By December 1944, 37 soldier mechanics were assigned; in March 1945, there were 51 enlisted men and 22 civilians, of whom 26 enlisted men and 8 civilians served as mechanics, and the rest worked at lubricating and tire repairing, and in the body shop, welding shop, and filling station. In April 1946, the staff included 97 civilians and 53 enlisted men.

Most equipment received after December 1944 was procured from excess at Hanford, Oak Ridge, etc., and its mechanical condition was only fair. Many items were "dead lined" upon arrival, or classified mechanically harmful or dangerous to operate until repaired, and the rest required constant maintenance. Very little new or rebuilt equipment was received, so it was necessary to buy spare parts on the open market, in order to continue operations. Ordnance sources were canvassed with little success. Only small items were purchased on the open market for stock, the rest were bought on emergency requisitions. To keep the equipment operating and the Project functioning, the District Office granted permission for these local purchases. The distance from the supply points made the Project entirely dependent upon motor transportation for supplies, subsistence, construction material, and personnel.

Miscellaneous Subjects

Congressional Inquiries. Two inquiries from United States Senators asked for information pertaining, respectively, to alleged destruction of serviceable Government property and to discrimination against housing of employees of Spanish ancestry. A careful investigation revealed that the claimed destruction of Government property involved old, obsolete furniture that would have cost the Government more to renovate than to replace. The total estimated cost was $2,680, and some parts were salvaged or reclaimed. Proper Army procedure for ridding the Government of these useless articles was followed. The individual who initiated this inquiry, by complaint to Senator Dennis Chavez, was interviewed by proper authority. It was found that he had never visited the Project, and that he refused to identify the individual who had wrongly informed him.

The inquiry about housing discrimination was addressed to the Secretary of War by Senator Hatch and referred to General Groves for investigation. The investigation revealed that an influx of professional employees of the University of California had forced new construction of low-cost housing units, and that several families of janitors and carpenters, of both Anglo and Spanish ancestry, had been moved to the new units to make room for these professional employees. No racial discrimination was involved, because many more Anglo employees were living beyond the Project boundaries and forced to commute as far as 50 miles to work. A result of the inquiry was to increase the size of some of the new housing units, to accommodate the abnormal size of several of the families involved.

Town Council. The Town Council was formed in August 1943, by joint action of the Commanding Officer and the Project Director. It was established as an advisory council of six members, elected for six-month terms, from adult Project residents. It was responsible for discussing problems affecting community welfare, and of communicating its findings and recommendations to the Administration. It had the wider responsibility of maintaining a spirit of community cooperation directed toward a single objective, the efficiency and success of the Project. These aims were formalized in a constitution proposed by the Council and accepted by the Commanding Officer in April 1944.

The Town Council discussed and made recommendations on many matters affecting community administration and welfare. Most were minor administrative problems, such as PX and mess hours, housing-assignment policy, traffic control, and licensing and control of dogs. The Council sponsored or supported a number of community activities, such as Bond and War Fund drives, fire prevention, organized recreation, and Victory gardens.

The most important items involving some expenditure for which Council recommendations were at least partly responsible were: the help-yourself laundries, extra storage space for apartment dwellers, a children's play area and other children's recreational facilities, increasing floor space in one-bedroom McKee apartments beyond original plans, lowering rentals for efficiency apartments, and adjusting rentals to make those based on University of California monthly salaries comparable to those for Civil Service employees working for the Army Administration.

Labor Recruiting. In January 1943, U.S. Engineer Office, P.O. Box 1539, under the Manhattan District, was established in the Bishop Building, Santa Fe, New Mexico. For security reasons, the Santa Fe Area Engineer of the Albuquerque District personally recruited and interviewed all key personnel for the service installation, usually employees with a very good service record within the War Department, who could be made available for transfer. The main source of recruitment of skilled and unskilled hourly employees was through the U.S. Civil Service Commission and the U.S. Employment Service (War Manpower Commission).

Construction employees were paid in accordance with the Davis-Bacon Act of August 1935 for this area. Hourly maintenance employees were paid in accordance with the Wage Schedule for the New Mexico Area, approved in October 1944 by the Wage Administration Board, Washington, D.C.

Because of the rapid growth of the Project, and the increase to approximately 1700 Civil Service Employees in March 1945, the Personnel Section was increased to 20 employees. A peak of approximately 2600 Civil Service employees, reached in March 1946, was reduced to 419 in December through transfer to the Zia Company, which had assumed responsibility in late April.

The only labor problem with Civil Service employees was the opposition of organized to nonorganized labor. In November 1944, a representative of the Chief of Engineers, Washington, D.C.; the Director of the 13th U.S. Civil Service Region, Denver, Colorado; and an Officer from the Project met with the local Union officials in Santa Fe. The officer representing the Corps of Engineers informed all concerned that the Project was not a "closed shop" and that the Corps of Engineers would not discriminate between union and nonunion employees and would hire any qualified personnel certified by the U.S. Civil Service Commission. The labor situation thereafter was satisfactory.

Safety Record. A Safety Committee handled all safety problems until January 1945, when the Administrative Board decided to secure a full-time Safety Engineer. In March, a Safety Division was organized and a safety program was formulated to deal with both Post and University of California safety problems. Then in May of the same year, a distinct division was made between these two.

From the beginning of the Project until the end of 1946, there were 26 fatalities from the following causes.

7 from construction accidents
6 from traffic accidents
2 from falls
1 from drowning
2 from radiation exposure
2 from accidental shooting
1 from a smudge-pot explosion
2 from self-administered poison
3 from accidentally drinking ethylene glycol

The Award of Honor was presented to the Manhattan Engineer District by the National Safety Council, for "Distinguished Service to Safety." The accident experience of Project Y was part of the Manhattan Engineer District.

OPA Prices and Regulations. The question of retail price was checked with the Office of Price Administration, Albuquerque, New Mexico, which expressly stated that War Department stores and commissaries were exempt from retail regulations of the OPA. Because of this ruling, the Commissary, mess halls, and Post Exchange did not enforce OPA ceiling prices effective in this area. However, with few exceptions, retail prices were considerably lower than those in Santa Fe.

One apparent exception was milk. However, the contractor chosen to supply milk to the Project was the only one who could meet the purity standards and bacillus specifications set by the Project Veterinarian. Investigation also showed that no other supplier could deliver the necessary quantities of high-quality milk. So, although the price of bottled milk on the reservation was 6 to 10% higher than the price in Santa Fe, the difference could readily be attributed to the difference in butter-fat content and the adherence to purity standards.

Disposition of Funds. Originally all monies from commissary sales, furnishing of subsistence, quarters, and utilities, and all other income accruing to the Government, were forwarded to the Finance Officer, U.S. Army, for deposit to the credit of the Treasurer of the United States. In February 1944, the contract with the University of California was modified, directing that all monies collected be forwarded to Lt. Col. S. L. Stewart, Contracting Officer, to be credited against the contract cost. In the Fall of 1946 this procedure was again altered and transmittal was made to the Zia Company. Figures furnished by the Fiscal Office showed the following transmittals:

Funds Transmitted to University of California

1944	$ 588,587.11
1945	1,277,631.84
1946	1,833,188.10

Funds Transmitted to the Zia Company

1946	$ 11,839.85

Claims Against the Government. The only claims filed against the Government were comparatively small ones, such as those resulting from automobile accidents and from blast effects, which were settled in a routine manner. After Frijoles Lodge had been released by the Manhattan District as extra housing, a claim for damage was entered against the Government for injury to property. This and other similar small claims were settled by the University of California.

Legal and Patents. In a community the size of Los Alamos there were naturally a number of situations in which legal counsel was required, not only in the official project functions but also in the personal affairs of the residents and military personnel. Because of the isolation of the community, local attorneys were not readily available and, further, the desire for maximum security suggested the use of resident counsel that might be available. Captain Ralph Carlisle Smith was assigned to the Technical Area in June 1943, as a special representative of the Office of Scientific Research and Development. In addition to the primary duty of Patent Advisor, Captain Smith was placed in charge of the Legal Section of the Post Headquarters. Eventually other attorneys joined the Project and shared the legal duties. Among the responsibilities of this group was the contacting of local authorities in State organizations, to provide for integration of the community without conflict with local regulations, while maintaining security.

Population Figures. No official census was attempted at Los Alamos until April 1946, as the total population was considered highly classified information. The first large group of employees, approximately 1500 Sundt construction workers, came in January 1943. By the 1st of July 1943, it was thought that most of the construction had been completed, and the force was reduced to about 1,000. Then came another rush of development when 1800 workers were added to Sundt's payroll. Morgan and Sons had almost 400 employees during the four months their contract was in effect. Statistics on McKee's employees fluctuate constantly, reflecting periods of construction activity. The University of California and Civil Service employees showed a rapid increase at first, then a gradual climb. At the beginning of the Project, practically all dependents were working, as housing was provided only for employees. As housing became available, the ratio of dependents increased, mainly because military personnel without resident dependents were replaced by civilian personnel with families. The total population as of January 1943, was approximately 1500; by the end of 1944, it had reached 5675. There was a sharp increase during 1945 to an estimated total of 8200. This growth continued throughout 1946, and the population reached an estimated total of 10,000.

VI. MILITARY ORGANIZATION AND PERSONNEL

Administration

Lt. Colonel J. M. Harman was originally assigned as Commanding Officer at this station and was sent out to set up his organization primarily in a service capacity, to take care of the feeding, housing, general comfort, and welfare of the University of California personnel. Lt. Colonel Harman set up the original office in Room Number 8 of the Bishop Building, in Santa Fe, in January 1943. Letters from General Groves (see next page) outlined the personnel and supply requirements for this Project, and gave the Commanding Officer of the new military post adequate power to perform his assignment.

The original plan was to have approximately six officers and a minimum number of civilians in key positions, with a group of WAC's and enlisted men to serve in clerical capacities. This arrangement was soon made obsolete by the development of the Project and the immediate need for a larger administrative staff. Each period of growth resulted in organization changes.

Colonel Harman, who was promoted in February 1943, remained Commanding Officer until May when he was succeeded by Lt. Colonel Whitney Ashbridge. Lt. Colonel Ashbridge commanded the Post until October 1944, at which time Colonel G. R. Tyler became Commanding Officer. He, in turn, was succeeded by Colonel L. E. Seeman in November 1945. Colonel Seeman was in Command until September 1946, when Colonel Herbert C. Gee was assigned that duty.

In addition to the Army staff, key civilians were appointed to act as supervisors in various sections. The Civilian Personnel Section was organized under J. P. Adams; P. A. Curran assisted Captain E. A. White in operating the Procurement Section; Town Management was the responsibility of F. W. Grefe and R. B. Osborne; the Commissary was managed by M. H. Gurley, and the Warehouse and Property Manager was V. R. Lefler.

From the small group of 32 Civil Service employees on the Post in February 1943, the strength grew to a peak of 2600 in March 1946. In April 1946 approximately 1800 employees were transferred to the newly organized Zia Company.

Military Personnel

In addition to the Headquarters staff, three troop units were originally assigned to the Post, and a fourth was initiated in October 1943. They were the Military Police Detachment, the Provisional Engineer Detachment, the Provisional WAC Detachment, and the Special Engineer Detachment.

The MP Detachment, 4817 Service Unit, started in April 1943 with seven officers and 196 enlisted men, plus a veterinary and a medical officer. The MP officers were Captain A. L. Cernaghan, 1st Lieut. C. E. Day, 1st Lieut. E. V. Hughie, 1st Lieut M. Wroe, 2nd Lieut. R. M. Cassidy, 2nd Lieut. H. C. Bush, and 2nd Lieut. J. F. Vollmer; 1st Lieut. J. J. Horowitz was the Medical Officer, and 1st Lieut. R. E. Thompsett was the Veterinary. The principal function of this detachment was to serve as security guard for the Project. As the Project grew, many additional guard posts were necessarily established, and the unit was authorized a new strength of nine officers and 486 enlisted men. By the end of December 1946, there were approximately 500 enlisted men assigned to this Detachment.

The Provisional Engineer Detachment with one officer, 1st Lieut. Clinton A. Nash, and 42 enlisted men reported to Colonel Harman in April 1943. It was purely a service organization with all personnel picked to fill specified jobs. They operated the power plant, steam plant, motor pool and garage, and mess halls and repaired and maintained buildings and roads. This unit was increased to 256 men and two officers, 1st Lieut. C. A. Nash and 2nd Lieut. T. F. Huene, to fill the needs of the growing community, by taking over positions in the Commissary, Post Exchange, and Post Engineer Office which could not be filled by qualified civilians. Later still, this detachment was increased to 465 men because of additional mess halls, enlargement of power and steam plants, and the larger number of motor vehicles. A small sawmill also was set up and operated by this unit. Both the MP and Provisional Engineer Detachments were subsistent on garrison rations, secured from the Quartermaster at Bruns General Hospital, Santa Fe. The MP Detachment operated the mess hall and supply room for both

February 22, 1943

SUBJECT: Organization and Assignment of Military Organizations, Los Alamos, New Mexico.

TO: The Commanding General, Services of Supply.

1. The extremely critical and secret nature of the operations to be carried on at Los Alamos to be activated on April 1, 1943 requires that it be protected and guarded by a special military organization fully subject to military control and suitable for periodic transfer outside the continental limits of the United States and for control of their activities after being so transferred. The unit should not be sent to a theater where members would be subject to capture or interrogation by the Germans.

2. It is requested that a special Military Police Company (reinforced) be organized and assigned to Los Alamos; that the Provost Marshall General, the Surgeon General and the Chief of Engineers be directed to participate in the organization, initial training, and equipping of the organization; that allotments of Medical, Military Police and Engineer troops and equipment be established for this project; that the withdrawal of trained personnel from existing tactical units be authorized to make up the initial complement; that only personnel be assigned who are of unquestioned loyalty and are suitable for overseas duty exclusive of the European and African Theaters.

3. The training of personnel should be similar to that normally given guard units. It would be advantageous if a large portion of the personnel were accustomed to horses.

4. The organization should be of the following magnitude:

Military Police	190	Officers and men (including company headquarters)
Medical and Veterinary	9	Officers and men
Engineer	45	Officers and men
Total	245	Officers and men

The details of the organization and equipment will be developed with the Provost Marshall General, the Surgeon General and the Chief of Engineers as soon as the organization is authorized.

-1-

5. The Units, trained and equipped, should arrive at Santa Fe, New Mexico between April 1 and April 10, 1943.

/s/ L. R. Groves
L. R. GROVES
Brigadier General, C.E.

(S E A L)

WAR DEPARTMENT
SERVICES OF SUPPLY
A P P R O V E D
MAR 2, 1943
FOR THE COMMANDING GENERAL
/s/ W. D. STYER
CHIEF OF STAFF

A TRUE COPY:

Edith C. Truslow

EDITH C. TRUSLOW
2d Lieut. TC

-2-

organizations. The Provisional Engineer Detachment had 302 enlisted men assigned to it when it was inactivated on July 1, 1946.

The WAAC Detachment was activated in April 1943, as the First Provisional WAAC Detachment. Its strength was the Commanding Officer (3rd Officer Helen E. Mulvihill) and six auxiliaries. These seven reported to Los Alamos in April 1943. In June, four more auxiliaries joined the unit, which grew slowly but steadily. The entire organization of two officers, 2nd Lieutenants Mulvihill and Creighton, and 43 enlisted women, was sworn into the Army of the United States by Lt. Colonel Ashbridge, in August 1943.

The working program for WAC's was not well defined at first, and all were put on basic jobs, although many of them had technical qualifications. However, as they proved their ability, they were placed in practically every department on the Project. Several were engaged in scientific research, and many were in positions handling highly classified material. They were also librarians, clerks, telephone operators, cooks, and drivers, and approximately 20 medical WAC's served as hospital technicians. In August 1945, the peak month, the WAC Detachment numbered approximately 260, including assigned and attached personnel. In June 1946 it had decreased to 84 because of the Army's demobilization program. The detachment was inactivated in October 1946.

The Engineer, MP, and WAC Detachments were all parts of a Service Command Unit, their specific title being 4817th Service Command Unit, 8th Service Command Detachment.

The Special Engineer Detachment was established in October 1943 as a unit of 9812 Technical Service Unit, directly under the District Engineer, and did not come under the jurisdiction of the Eighth Service Command. This unit was formed to retain University of California employees who were drafted and to supply trained technical personnel for positions that could not be filled by civilians. Most of the enlisted men in this unit came directly from ASTP schools and were selected on the basis of previous background and education. They were mostly mechanical, electrical, and chemical engineers, or men of similar scientific backgrounds. Enlisted personnel of this type proved extremely valuable, and the unit grew rapidly to 1823 men by October 1945. Demobilization cut this to a low of only 403 by June 1946, but Regular Army replacements were used to build the detachment up again to about 800 men in December 1946.

Special efforts were made to obtain officers who had outstanding experience and training and an unimpeachable security background. However, as was to be expected over a period of years, there was a substantial turnover of officers assigned to the station. Some men were assigned for a special duty, and when this responsibility was completed they were released for reassignment in the Manhattan Project or elsewhere. Other officers were transferred, sometimes at their own request, because they were unable to adapt themselves to the unusual conditions at the Project, or sometimes because their special training was not being used to best advantage. A very few officers stayed at Los Alamos from the early stages in 1943 through V-J Day and later.

Considerable hostility developed between the Technical Area civilian workers and the workers, civilian and military, in the Post Administration. The original arrangement of the camp contributed materially. In addition to a fence around the main Project Site, there was a fence around the Technical Area. However, the fence that caused the greatest friction was one that cut across the main site at the Post Headquarters. This fence separated the Military Police and Provisional Engineer Detachments and construction camp workers from the rest of the population. During the day, individuals were free to pass through the two gates in this fence. However, at about six in the evening one gate was closed and a military policeman stood guard at the other. Members of the general civilian area were permitted to go to Theater No. 1 in the Military Area, but military personnel and construction workers were not permitted to enter the general civilian area without an invitation. Furthermore, the Post Exchange in the civilian area received more extensive equipment and supplies. Although the segregation did not last long, a natural resentment arose from which the Project never recovered.

The original group of enlisted men serving at Los Alamos had volunteered for overseas duty and were resentful of their assignment, particularly when they observed many young men who were not in uniform. Most of the officers and none of the enlisted men of the original detachment were permitted to know specifically the nature and importance of the work. Consequently the Technical Area workers, both civilian and military, were called "phonies," "technical-area jerks," and "longhairs" by a substantial proportion of the Post organization. This situation was not improved by the fact that Technical-Area civilian workers could have housing for dependents, whereas military personnel could not. In some cases menial tasks for the civilian population were assigned to the enlisted personnel, causing further dissatisfaction. Other irritating items included the lodging of the detachment officers in semibarracks and the refusal to permit them to form a regulation club. Actually, the Army and Navy officers did form an association known as the Officers' Lounge, but were careful to invite civilians, to avoid charges of discrimination.

Few of the military, naturally, were at Los Alamos by their own decision, and most of them were unhappy with their lot. Furthermore, the military were bound by regulations that did not always permit them to accede to the requests of the civilians. The refusals by the military were not always gracious, and, as a result, a considerable portion of the business was done at arm's length.

Promotions at Los Alamos for a substantial period were very slow, and at times the privilege of furloughs and leaves was suspended for security reasons. Recreational facilities were inadequate. Censorship of mail materially reduced contact with the outer world.

With the addition of the Special Engineer Detachment, morale became even worse. That unit's Table of Organization permitted all members to be noncommissioned officers with the consequent benefits. One-third of the men were staff sergeants and higher, one-third duty sergeants, and one-third corporals or the equivalent in technician ratings. Many were completely without basic training, but received their ratings on the basis of their Technical Area assignments. The civilian scientists became interested in these military co-workers and entered into the matter of promotions, details, and punishment extensively. This apparent infringement on the military prerogative was not accepted gracefully.

Even before the nature of the project was published, many enlisted men tried to obtain transfers to the Special Engineer Detachment, and in a few cases this was done. Even more important, however, was the effect on morale of the news that Los Alamos had produced the atomic bomb that was dropped on Hiroshima. The men who were most bitter about the assignment and most resentful of the civilians suddenly were an intimate and vital part of the Los Alamos laboratory that had performed this exceptional job.

APPENDIX

EARLY S-SITE EXPERIENCES

by

Edward Wilder, Jr.

The following is an account of the early days at S Site, as I remember them. I have made no effort to confirm the dates, but relative events are probably correct. I hope that this account will enable future writers to have a feeling for some of the conditions under which the first atomic bomb was developed and built.

My introduction to the Manhattan Engineer District was through the Navy. I was an enlisted man (draftee) and had taken the routine aptitude officer candidate tests. As a result of these tests, I was commissioned a Lieutenant (jg). At about this time (so rumor had it at Oak Ridge) Eastman at Oak Ridge was behind in their production. Their explanation was that they were short of technical help. So, to fill this need, the next 150 naval officer candidates with engineering degrees being inducted were sent to Oak Ridge. Most of these men had absolutely no military training. They were sworn into the Navy, bought a uniform, and reported to Oak Ridge all in a few days. This created situations that could be the subject of a good comedy.

Some of these men were very unhappy at Oak Ridge. They had joined the Navy to fight a war, not to work in a factory. Also some of them had influence in Washington. The result of their agitation was that in March, 1944, the unit was disbanded. About 15 of the men were sent to Los Alamos. I was selected by Commander Bradbury, who interviewed us at Oak Ridge, and was told that I would work on the development of procedures for machining a material that had never been machined before. This material, I learned later, was high explosives.

I was not told where I was being sent, except that it was in the Southwestern part of the U.S. My orders instructed me to report to 109 East Palace Avenue, Santa Fe, New Mexico. I was told that the Santa Fe address was not my duty station. So on March 15, 1944, Ensign W. A. Wilson and I arrived at 109 East Palace Avenue and reported to Mrs. Dorothy McKibben and were told how to get to Los Alamos.

Junior officers were not allowed to have their families with them, but they were housed in a very nice (for Los Alamos) dormitory called the Super Dorm. Meals were usually eaten at Fuller Lodge, although various mess halls were available. Food was cheap and very good.

The area south of the town site was closed to nontechnical traffic during work hours. The road through this area was what is now Anchor Ranch Road and West Road. The guard stations were in Los Alamos Canyon where the skating rink is now, and at State Road 4. When the road was open to the Los Alamos public (holders of Post passes), the occupants of each car were given a pass that showed the time and the number of people, at one guard station, and they had 20 minutes in which to check out at the other guard station.

Letters mailed in Los Alamos were put in the mail box unsealed. They were read and, if approved, were sealed and sent on by the censor; if they were not approved they were returned to the sender. There was to be no evidence of censorship at Los Alamos to the outside world. Incoming mail was opened and read.

When I arrived, Los Alamos was expanding. A story then current was that it was originally thought that 300 of the best scientists, engineers, and technicians would be brought in, locked up, and told to build the bomb. In 1944 we believed that we occasionally were trailed by the FBI when we left the Hill. When we traveled on

Project business we were instructed to avoid conversation with strangers and if forced by circumstances to discuss where we were from, to say Washington, D.C.

We were discouraged from making friends in Santa Fe. As a junior officer, I did not qualify for housing on the Hill. When my wife and a friend's wife decided to visit us in June 1944, I was instructed to tell them that they could not live closer than 100 miles from here. Albuquerque was OK, but Santa Fe was definitely out. They came anyway, and lived in a motel for several weeks. This was not satisfactory, so we brought them to Bandelier. We lived there until the end of August 1945 when the war was over and housing became available. Their living so close to Los Alamos was definitely against the rules, but because there was little or no tourist traffic, we were not forced to move by the Project officials. To keep Security from noticing us, we always left the canyon on Sunday.

The weather in 1944-46 was different from the 1970 variety. The winters had more snow, and the summers more rain. The snow created problems because we had almost no snow-removal equipment. The summer rains were accompanied by violent thunder and lightning storms. These storms were so frequent and violent that shelters were constructed where the workers could go to get away from the explosive processing.

When the Los Alamos project was first established, the local public milk supply was inadequate and below standard. One of the local dairies was soon selling milk here. The story was that the Government had up-graded the dairy so that its product would be acceptable. It was thought that the health hazard associated with working with TNT could be reduced if those concerned drank lots of milk, so free milk was provided for meals and mid-shift breaks.

Most of the roads were gravel, some very rough. Once when General Groves visited Los Alamos, Kistiakowsky took him to S Site in a jeep that had the springs made inoperative by wooden blocks under them. As a result of that trip, the roads over which HE was moved were improved.

I was assigned to Group X3 under Major Jerome (Jerry) O. Ackerman. Our responsibility was to develop and manufacture the high explosives system for the implosion bomb. This was done at S Site. It was on the south side as far away as possible from the rest of the Project. We believed that this was because of the danger involved in what we were doing. The personnel at S Site was almost completely military. We also believed that this was because the work was too dangerous for civilians.

We worked a nominal 8 to 5 shift, but, in practice, everyone worked much more than this. The plant worked a 24-h day, six days a week. The product in the early days of the Project was used to study materials under conditions of high pressure and shock, as well as to develop the explosive lens that would implode the nuclear components of the bomb. The operation consisted of melting HE and pouring it into molds whose shape was determined by theoretical calculations.

The men who conceived the idea of using explosives to implode the bomb underestimated the difficulty of doing this. As a result, the first facilities built were completely inadequate. This is no adverse reflection on these men, who were the foremost men in this field. Rather, it is an indication of the trouble that can be encountered in applying a familiar product to a new use. I do not know whether the planners of S Site anticipated having to build additional facilities, but there was continuous planning and construction of new buildings until just before Trinity day.

Notably absent from the first S-Site buildings was any where machining could be done. The first HE castings were worked with hand tools, saws, rasps, and planes, to a template. This was done on boxes and makeshift stands on the floor of Building TA-16-26. One of the first memories I have of S Site is that of a man sawing a big piece of HE on a Comp. B box. He had his knee on it just as if it were a block of wood. His left hand held a hose and directed a stream of water onto the saw which he held in his right hand. I believe that explosives other than Comp. B and TNT were processed before I arrived. I do not remember working with anything but Comp. B, TNT, and Baratol. Also it seems that we were making more full-size castings than scale models.

As it seems now, the information and experience learned in Building 26 indicated that more scale-model HE assemblies would be required. So the next unit of construction was built for this purpose. It was called at that time the 30 line. Under present designation, these buildings were TA-16-42, Casting; TA-16-43, Machining; TA-16-44, Physical Inspection; TA-16-45, X Ray; and TA-16-46, HE Storage for X Ray. Shortly after these were built, TA-16-48 was constructed as a gamma-graph facility. The service area was also being expanded during this time. The number and function of these buildings indicate the magnitude of the problem of making the explosive component of the bomb.

The first studies of the liquid-solid interface of HE as it solidifies in a mold were made by pouring out the molten material at different time intervals after the mold was filled. Later, in Building 26, thermocouples mounted on a balsa-wood "christmas tree" were placed in molds, and the cooling pattern was studied in that way.

The kettles used in Building 42 were stainless steel candy kettles, jacketed and steam heated. The agitator was driven by an air motor. The molten explosive was poured from the kettle into a rubber bucket and from the bucket into the molds. The molds were steel weldments of the shape desired, a pentagon or hexagon. Small-diameter tubes were fastened to the inside of the steel for circulation of tempering water. The mold was finished with Cerrotru, a low-melting casting alloy, around a master shape supported in the steel weldment. This Cerrotru covered the tempering coils and produced the mold that was used in producing the explosive shapes. The mold had no inner or outer radius surface. The inner radius was produced by an insert, called a toadstool. The outer radius was produced by a surface attached to the mold cover. If the piece being made was of Comp. B, the outer surface was spherical; if it was Baratol, the outer surface was shaped accordingly.

In Building 42 we made the first effort to control the cooling pattern of the explosives as they solidified in the molds, by the use of steam heat on the risers and water of selectable temperature on the body of the mold. Another device for improving the casting quality by reducing segregation was to use an air-motor-driven stirrer in the casting during the cooling cycle. One of the duties of the casting-room attendants was to raise these stirrers as the material solidified.

Here we should acknowledge the dept that development of the bomb owes to self-adhesive tape. In my opinion, development of the explosive component of the bomb was greatly facilitated by the use of sticky tape. Special molds and risers were made of cardboard and tape. Things were repaired with tape. It seemed that it was used almost everywhere. I believe that a Navy junior officer named Glenn Greening spent most of his time working with Minnesota Mining & Manufacturing Company in developing tapes for different purposes.

It was in Building 42 that I saw a very dramatic demonstration of the inherent safety of the explosives we were processing. In the process we were using at that time, it was essential that the mold and riser unit be partially disassembled before complete solidification. Otherwise, there was no easy way to take them apart. On this day a mold and riser had solidified, and one of the operators was attempting to remove the riser. He was using a bronze screwdriver and a rawhide mallet to chip the HE away so that the riser could be removed. This was standard procedure, but on this day for some unknown reason there was an explosion under the screwdriver. It was a small explosion involving only one crystal and did not propagate to the rest of the explosive.

In these times there was no safety organization. After VJ Day, various safety men appeared, but they were uncertain and unknowledgeable, and accomplished little. The first real safety man at Los Alamos, in my opinion, was Roy Reider.

After casting, the HE was taken by hand truck to Building 43 to be machined. The equipment consisted of one K&T milling machine located in a barricade corner. The other machines were wood-working-type Delta drill presses. On liners and inner charges, the risers were machined off by using fly cutters finished to the correct curvature. Comp. B was machined under a stream of water. Baratol was also machined with fly cutters, but the blades had a long spiral shape. Baratol was initially machined dry. The reason given was that water would dissolve the barium nitrate. It was only after I put a piece of Baratol in a sink and let water drip on it overnight to prove that this was not important, that Baratol was machined wet. The K&T machine was used to a limited degree in milling the flats on lenses. Its principal use was in shaping rectangular blocks of HE to definite dimensions for theoretical studies.

The first machinist at S Site was Ernest Ritchie. He was a little man who weighed about 110 pounds. I remember his lifting and machining full-scale lenses that weighed about 125 pounds.

Building 44 was used for the dimensional inspection of the castings. The equipment used was surface plates, dial indicators, and conventional dimensional standards. Building 45 was used to x ray castings. Portable 150- and 220-kV x-ray machines and lead screens were used. The building also had a darkroom where the film was processed. Building 46 was used as a rest house for castings during the dimensional and x-ray inspection. Building 48 was built soon after 1946. It housed a radium source used in gamma-graphing large or high-density objects.

Building 55 housed the barium nitrate grinding machinery. This was one micropulverizer, a small, high-speed hammer mill. I remember an operator's being knocked down when he touched the aluminum can, which received the ground nitrate from the pulverizer, by the static charge on it. If there was a place where barium nitrate was ground before this I do not remember. If there wasn't, the material must have been used as received from the factory. If this were so, it is obvious now why they found that grinding was necessary.

Building TA-16-81 was used to dry nitrocellulose. It was really two buildings. A small one housed a steam unit heater that blew its hot air through a large pipe to the main building where the nitrocellulose was spread out in trays to dry.

All explosive operations produced great quantities of scrap. This was collected daily and burned in an area where the east end of Building TA-16-260 is now. For the burn, the material was spread out in a single layer. Big pieces (125 lb) were broken up into several smaller ones by hitting them with a heavy rubber mallet. The blows had to be very hard, and one man who did this was fat and always short of breath. Once, as he was breaking up some HE, a piece flew in his mouth and he swallowed it. We did not know how this would affect him, but apparently he suffered no ill effects.

Several times the explosive detonated instead of burning. The man in charge of the burning ground, and who ignited the HE, could speak clearly under normal conditions, but when he was excited he stuttered. Once when the burning ground exploded with a terrific bang, I hurried there to see if anyone was hurt. I met him driving away from the burning ground. He stopped me and said "everything is all right, the burning ground just blew up," but it took about three minutes for him to say it.

Another time as we were spreading some cordite on the ground preparing to burn it, it caught fire. I had always heard that a man in real danger will act impulsively to save himself, and this is what we did. Every man ran in the direction he was facing when the fire started. Some ran farther than others. I ran only a short distance and threw myself into a ditch, completely ignoring the stickers and tumble weeds.

Building 27 was built during 1945 and was used to make full-scale castings. The kettles were larger, and the temperature of the water used to cool the molds could be varied as required. Buildings 30, 31, 32, 33, and 34 were built at the same time. They were used to machine the Baratol and Comp. B castings made in Building 27. Buildings 30 and 34 were magazines, and Buildings 31, 32, and 33 were machining bays. The machines used were 48-in. Fosdick radial-arm drills. We were told that they were "lend-lease" machines taken off a ship destined for Russia. By the time these buildings were designed, we were getting more safety conscious. They were provided with a remote-control building located behind the operating bay and connected with it by a 12-in.-diam pipe for viewing and a smaller pipe for controls. These pipes were never used.

Shortly after these buildings were built, it became desirable to machine all surfaces of the HE material. This was done in Buildings 94, 95, 96,97, and 98. Building 94 was a central equipment room serving the four surrounding machining bays. As I remember, this facility was designed and built in 30 days. Building 98 was equipped for remote operation and viewing. The viewing was through a submarine periscope mounted horizontally.

At about this time, we began to have real trouble in manufacturing full-scale lenses. Baratol components as made, inspected, and accepted would not, when put in a mold for overcasting, fit the mold. I remember working many nights trying to find out what was happening to our process. We finally realized that the Baratol was changing dimensions. This is known and understood now. Then it was new, but we were able to develop enough of an understanding to produce acceptable lenses.

The buildings that were then and now called V Site, TA-16-515, 516, 517, 518, 519, and 520, were a separate operation not under GMX-3. At V Site there was a large mechanical shaker on which were run preliminary tests on the bomb. This activity was terminated some time before Buildings 30 - 34 were built. The S-Site activities were constantly expanding, so we moved into V Site.

The lens and inner charges were large and heavy, and the explosive material was rather fragile. We used three methods to protect the HE from chipping. The finished casting was sprayed with the best "Bar Top" varnish available, a thin layer of felt was glued to one of two mating spherical surfaces, and blotting paper was glued to the sides. This was done in Buildings 519 and 520.

As things developed and it looked as if a bomb would really be built, the practice assembly of the HE components was started. This was first done in Gamma building in the main Tech Area. As I said above, we were not bothered by Safety people. The assemblies were made on a floor area padded with wrestling mats. Sometimes the pieces did not fit very well. I remember someone hitting a piece of HE that was out of line, as hard as he could, with a heavy rubber mallet. The Trinity bomb was assembled in Building 516. At the end of the day, when the HE components were all put together, I thought that it should be guarded during the night. It was with great difficulty that I convinced people that this should be done.

The Trinity test was conducted on July 16, 1945. Now we had to make a bomb that could be put together in an airplane. This was done. The Hiroshima bomb was used on August 6, and the Nagasaki bomb on August 9, 1945.

After Japan surrendered, things slowed down a great deal. The most notable event after VJ Day was in the winter of 1945 and 46 when the water supply to Los Alamos froze. When it was apparent that the pipe lines from Guaje and Los Alamos canyons could not be kept flowing, everyone who could possibly leave was encouraged to do so. This happened very quickly. I remember we set up a furlough-issuing line in the cafeteria. The Red Cross was there to lend money to those who did not have the cash. Water for the town and those people who stayed was trucked in. Every tank truck for hundreds of miles around was brought in and set to hauling water from the Rio Grande to Los Alamos. At this time, it was estimated it cost about 25 cents to flush a toilet. This was the low point in morale on the Hill. From there on things got better.